INTRODUCTORY THERMODYNAMICS

Pierre Infelta

Swiss Federal Institute of Technology
Lausanne, Switzerland

BrownWalker Press

Boca Raton, Florida

INTRODUCTORY THERMODYNAMICS

Copyright © 2004, Pierre Infelta.

All rights reserved.

No part of this book may be reproduced or translated without the prior written permission of the copyright owners, except as permitted by law.

BrownWalker Press – 2004

www.brownwalker.com
1-58112-416-3 (paperback)
1-58112-421-X (ebook)

Library of Congress Cataloging-in-Publication Data

Infelta, Pierre, 1942-
   Introductory thermodynamics / Pierre Infelta.
       p. cm.
   Includes bibliographical references and index.
   ISBN 1-58112-416-3 (alk. paper)
1.   Thermodynamics. I. Title.

QD504.I54 2004
542'.369--dc22   2003027840

Foreword

In numerous fields, a certain level of understanding of thermodynamics is a necessity, yet the amount of time imparted to this end is often very scanty. Hence, the need for a simple, compact and easy to read text, intended for students beginning in the field.

This textbook provides a very concise yet clear presentation of classical thermodynamics in 178 pages. I include many examples (111 actually) to provide instant illustrations and applications of the results obtained.

Being concise implies that the reader has to accept a number of properties without rigorous derivations, such as "Entropy is a state function". We all know, however, that a good understanding and logical derivations significantly decrease the amount of material that needs to be memorized. Whenever possible, I present many simple (often somewhat original) derivations with intermediate steps in the derivations. Equations, without clear statements giving the extent of their validity, are useless. I made a point to clearly mention, where needed, the conditions under which a relation is valid. I also include a very extensive index.

The *ebook* version is a very small file, instantly available for download. It is very useful for efficient searches, highlighting and writing notes and is enriched with colored figures.

I wish to express my thanks to the many colleagues, former colleagues and friends who have contributed so significantly to the quality of this book by their careful reading and their subsequent comments.

Pierre Infelta
School of Basic Sciences,
Swiss Federal Institute of Technology
Lausanne, Switzerland

Table of contents

1. **Thermodynamic Systems : Definitions** ---------- 1
   - 1.1 Interactions of Thermodynamic Systems with their Surroundings ---------- 1
   - 1.2 Equilibrium ---------- 1
   - 1.3 Thermal Reservoir. Heat Source ---------- 1
   - 1.4 Diathermal and Adiabatic Enclosures ---------- 2
   - 1.5 Zeroth Law of Thermodynamics ---------- 2
   - 1.6 Intensive and Extensive Variables. State Functions and State Variables ---------- 2
     - 1.6.1 Definitions ---------- 2
     - 1.6.2 Fundamental and Auxiliary State Variables or Functions ---------- 3
     - 1.6.3 Thermal Coefficients ---------- 4
   - 1.7 Change of a State Variable as the Result of a Thermodynamic Process ---------- 5
     - 1.7.1 General Process ---------- 5
     - 1.7.2 Cyclic Process ---------- 5
     - 1.7.3 Mathematical Characteristics of a State Function ---------- 6
   - 1.8 Reversible and Irreversible Processes ---------- 7
   - 1.9 Equation of State ---------- 8

2. **Work** ---------- 9
   - 2.1 Sign Convention for Energy Exchange ---------- 9
   - 2.2 Mechanical Work ---------- 9
     - 2.2.1 Definition ---------- 9
     - 2.2.2 Work and Volume Change ---------- 10
     - 2.2.3 Process at Constant External Pressure ---------- 11
     - 2.2.4 Work during an Isothermal (Reversible) Change of an Ideal Gas ---------- 11
   - 2.3 Remarks ---------- 12
   - 2.4 Electrical Work ---------- 12
   - 2.5 Various Forms of Energy ---------- 13
   - 2.6 Various Expressions for Work ---------- 13

3. **First Law of Thermodynamics** ---------- 15
   - 3.1 Introduction ---------- 15
   - 3.2 The Joule Experiment ---------- 15
   - 3.3 Internal Energy. First Law ---------- 15
     - 3.3.1 General Aspects. Expression of the First Law -- 15
     - 3.3.2 Closed System – Adiabatic Process ---------- 16
     - 3.3.3 Closed System – General Process. First Law ---- 16
     - 3.3.4 Closed System – Cyclic Process ---------- 16
     - 3.3.5 Infinitesimal Process ---------- 16

4. Second Law of Thermodynamics —————————— 17
- 4.1 Kelvin Formulation of the Second Law ——————————17
- 4.2 Carnot Cycle. Heat Engine. Thermodynamic Temperature ——————————————————————17
- 4.3 Entropy. Reversible and Irreversible Processes. Equilibrium——————————————————————————19
  - 4.3.1 Definition ————————————————————————19
  - 4.3.2 The Second Law, Entropy and Spontaneity of Processes——————————————————————— 20
  - 4.3.3 System in Contact with a Single Thermal Reservoir ———————————————————————21
  - 4.3.4 Clausius Inequality ———————————————— 22
  - 4.3.5 Differential Expression for the Internal Energy and Enthalpy of a Closed System———— 23
  - 4.3.6 Equilibrium Condition ——————————————— 24
  - 4.3.7 Thermal Equilibrium ———————————————— 24
  - 4.3.8 Isothermal (Reversible) Expansion of an Ideal Gas——————————————————————————— 26
  - 4.3.9 Monothermal Irreversible Expansion of an Ideal Gas ———————————————————————— 27
  - 4.3.10 Reversible Adiabatic Process of an Ideal Gas -- 29
- 4.4 Carnot Cycle and Entropy ———————————————— 30
- 4.5 Heat Engines, Refrigerators, Heat Pumps ——————————31
  - 4.5.1 Thermal Machines ————————————————————31
  - 4.5.2 Efficiency of an Engine————————————————— 32
  - 4.5.3 Performance of a Refrigerator —————————— 33
  - 4.5.4 Performance of a Heat Pump————————————— 34
- 4.6 Otto Cycle or Beau de Rochas Cycle ———————————— 34
- 4.7 Stirling Cycle —————————————————————————— 36

5. Auxiliary Functions : Enthalpy, Helmholtz Energy, Gibbs Energy ———————————————————————— 39
- 5.1 Introduction———————————————————————————— 39
- 5.2 Closed Systems ——————————————————————————— 40
  - 5.2.1 Constant Volume Process (Isochoric Process) -- 40
  - 5.2.2 Constant Pressure Process (Isobaric Process)-- 40
  - 5.2.3 Monobaric Process——————————————————————41
- 5.3 Characteristic Variables. Fundamental Equations. Open Systems. Systems with Chemical Reactions ———— 42
  - 5.3.1 Generalities ———————————————————————— 42
  - 5.3.2 Internal Energy —————————————————————— 42
  - 5.3.3 Enthalpy ——————————————————————————— 43
  - 5.3.4 Helmholtz Energy (Helmholtz Function, Free Energy) ———————————————————————————— 44
  - 5.3.5 Gibbs Energy (Gibbs Function, Free Enthalpy) - 44
  - 5.3.6 Chemical Potential. Summary ————————————— 45
- 5.4 Maxwell's Relations ———————————————————————— 47
- 5.5 Thermodynamic Equation of State ———————————————— 49
  - 5.5.1 General Case ———————————————————————— 49
  - 5.5.2 Equation of State for an Ideal Gas ——————————— 50

5.6 Properties of C_p and C_V -------- 51
  5.6.1 Relation between C_p and C_V -------- 51
  5.6.2 Variation of C_V with Volume and of C_p with Pressure -------- 52
5.7 Physical Meaning of the Auxiliary Functions -------- 53
  5.7.1 Helmholtz Energy (Helmholtz Function, Free Energy) -------- 53
  5.7.2 Differential Form -------- 55
  5.7.3 Gibbs Energy (Gibbs Function, Free Enthalpy) -- 55
  5.7.4 Differential Form -------- 56
  5.7.5 Spontaneous Evolution and Equilibrium Condition -------- 57

6. Mixtures and Pure Substances : Partial Molar Quantities and Molar Quantities --- 59
  6.1 Homogeneous Functions and their Properties -------- 59
  6.2 Extensive Variables -------- 60
  6.3 Intensive Variables -------- 60
  6.4 Explicit Expressions for Various Extensive Variables -------- 62
  6.5 Gibbs-Duhem Equation -------- 62
  6.6 Partial Molar Quantities -------- 63
  6.7 Molar Quantities. Pure Substances -------- 65
  6.8 Other Relations -------- 65

7. Thermodynamics of Gases -------- 69
  7.1 Pure Ideal Gas -------- 69
    7.1.1 Chemical Potential of a Pure Ideal Gas -------- 69
    7.1.2 Selection of the Standard State Pressure -------- 69
    7.1.3 Mathematical Expressions of other Thermodynamic Functions of Ideal Gases -------- 70
  7.2 Mixtures of Ideal Gases -------- 70
    7.2.1 Basic Properties. Ideal Gas Mixture. Dalton's Law -------- 70
    7.2.2 Chemical Potential of an Ideal Gas in an Ideal Gas Mixture -------- 72
    7.2.3 Mixing Properties -------- 74
    7.2.4 Irreversible Mixing of Two Ideal Gases -------- 75
  7.3 Pure Real Gases -------- 75
    7.3.1 Molecular Interactions in Real Gases -------- 75
    7.3.2 Chemical Potential of a Pure Real Gas -------- 76
    7.3.3 Fugacity Coefficient of a Pure Real Gas -------- 77
    7.3.4 The Virial Equation -------- 77
    7.3.5 The van der Waals Equation of State -------- 78
    7.3.6 Joule-Thomson Effect -------- 81
  7.4 Mixtures of Real Gases -------- 83
    7.4.1 Chemical Potential of a Real Gas in a Mixture -- 83
    7.4.2 Variables of Mixing for Real Gases -------- 84
  7.5 Ideal Mixtures of Gases -------- 86

8. Systems with Several Phases, No Chemical Reaction. Third Law of Thermodynamics —————— 89
- 8.1 Introduction —————————————————————— 89
- 8.2 Differential Expressions for State Functions ———— 89
- 8.3 Spontaneous Transfer of a Species from One Phase to Another One —————————————————— 90
- 8.4 The Phase Rule ————————————————————— 91
- 8.5 Equilibrium of Two Phases of a Pure Substance ——— 92
  - 8.5.1 Clapeyron Equations —————————————— 92
  - 8.5.2 Equilibrium between a Gaseous Phase and a Condensed Phase (Liquid or Solid) of a Pure Substance ————————————————————— 94
  - 8.5.3 Schematic Representation of some of the Thermodynamic Functions in the Vicinity of a Phase Change ——————————————————— 95
  - 8.5.4 Effect of the Pressure of an Insoluble Gas on the Vapor Pressure of a Liquid —————————— 95
  - 8.5.5 Effect of Temperature on the Latent Heat of Phase Change and on the Equilibrium Pressure — 97
- 8.6 Phase Diagram of a Pure Substance ————————— 98
- 8.7 Evaluation of Entropies ———————————————— 100
- 8.8 Third Law of Thermodynamics ————————————— 101
- 8.9 Implications of the Third Law ———————————— 101
  - 8.9.1 Heat Capacities ————————————————— 101
  - 8.9.2 Effect of Pressure and Volume on Entropy at 0 K ——————————————————————— 101
  - 8.9.3 Helmholtz Energy and Gibbs Energy at 0 K ——— 102

9. Energetics of Chemical Reactions ————————— 103
- 9.1 Introduction ————————————————————— 103
- 9.2 The Extent of Reaction ———————————————— 103
- 9.3 Variables of Reaction ————————————————— 104
  - 9.3.1 Gibbs Energy of Reaction ————————————— 104
  - 9.3.2 Other Variables of Reaction ———————————— 105
  - 9.3.3 Standard Variables of Reaction ————————— 105
  - 9.3.4 Standard Variables of Formation ————————— 107
- 9.4 Hess' Law ——————————————————————— 108
- 9.5 Kirchhoff's Equation ————————————————— 110
- 9.6 Effect of Temperature on the Entropy of Reaction and the Gibbs Energy of Reaction ————————————— 112
- 9.7 Conversion of Chemical Energy into Work ——————— 113
  - 9.7.1 Any Form of Work ————————————————— 113
  - 9.7.2 Work other than Work due to Volume Change — 114

10. Chemical Equilibria ———————————————— 117
- 10.1 Change in $G(\xi)$ with the Extent of Reaction ———— 117
  - 10.1.1 Expression for a Mixture of Reacting Ideal Gases ——————————————————————— 117
  - 10.1.2 Schematic Representation ———————————— 118
- 10.2 Spontaneous Reaction. Equilibrium ————————— 119

      10.2.1   Isothermal Isobaric System ---------------------- 119
      10.2.2   Isothermal Isochoric System ------------------- 120
      10.2.3   Adiabatic Isobaric System ----------------------- 121
      10.2.4   Adiabatic Isochoric System -------------------- 122
  10.3 Law of Mass Action for a Gas Mixture --------------- 123
      10.3.1   Standard Equilibrium Constant ----------------- 123
      10.3.2   Other Forms of the Law of Mass Action ------- 124
  10.4 Chemical Equilibrium in the Presence of Pure
          Condensed Phases ------------------------------------- 125
      10.4.1   Chemical Potential of a Pure Condensed Phase - 125
      10.4.2   Law of Mass Action for Heterogeneous
                Systems -- 126
  10.5 Independent Reactions --------------------------------- 128
      10.5.1   General Remarks --------------------------------- 128
      10.5.2   Number and Nature of Independent Reactions 128
      10.5.3   Equilibrium of Systems where Several
                Reactions can Take Place Simultaneously -------131
      10.5.4   Consequences on Equilibrium -------------------- 132
  10.6 Phase Rule for Systems with Chemical Reactions ---- 133
  10.7 Effect of Temperature on the Equilibrium Constant 134
  10.8 Displacement Laws of Equilibria ----------------------- 135
      10.8.1   Effect of Temperature ------------------------- 136
      10.8.2   Effect of Pressure ------------------------------ 136
      10.8.3   Effect of Volume ------------------------------- 137
      10.8.4   Effect of the Addition of an Inert Gas -------- 137

11. Perfect and Ideal Solutions ----------------- 139

  11.1 Basic Considerations ---------------------------------- 139
  11.2 Perfect Solution---141
      11.2.1   Isothermal Representation ----------------------141
      11.2.2   Isobaric Representation ------------------------ 143
  11.3 Mixing Properties of Ideal Solutions ----------------- 145
  11.4 Effect of Pressure and Temperature on Liquid Vapor
          Equilibria-- 146
  11.5 Lowering of the Freezing Temperature of a Solvent
          in the Presence of a Solute – Eutectic ---------------- 147
  11.6 Elevation of the Boiling Temperature of a Solvent in
          the Presence of a Non Volatile Solute ---------------- 150
  11.7 Osmotic Pressure --------------------------------------151

12. Non Ideal Solutions ----------------------- 153

  12.1 Introduction-- 153
  12.2 Variables and Excess Variables of Mixing ------------ 154
  12.3 Effect of Pressure and Temperature on the Activity
          Coefficient -- 155
  12.4 Standard State – Convention I for the Activity
          Coefficient -- 157
  12.5 Liquid – Vapor Equilibrium ---------------------------- 158
      12.5.1   Isothermal Diagram ----------------------------- 158
      12.5.2   Isobaric Diagram --------------------------------161

- 12.6 Standard State – Convention II for the Activity Coefficient --- 162
- 12.7 Liquid – Liquid Extraction --- 165
- 12.8 Other Composition Scales and Standard States --- 165
  - 12.8.1 Molality --- 165
  - 12.8.2 Concentration --- 167
- 12.9 Law of Mass Action for Liquid Phase Systems --- 168
- 12.10 Electrolytes --- 170
  - 12.10.1 General Considerations --- 170
  - 12.10.2 Chemical Potential of Ions in Solution --- 171
  - 12.10.3 Dissociation Equilibrium --- 172
  - 12.10.4 Hydrogen Ion Convention for Aqueous Solutions --- 173
  - 12.10.5 Electrode Potential --- 175

13. Bibliography --- 177
- 13.1 Textbooks --- 177
- 13.2 Handbooks and Tables --- 177
- 13.3 Articles --- 178

Index

1. Thermodynamic Systems : Definitions

1.1 Interactions of Thermodynamic Systems with their Surroundings

A thermodynamic system is a part of the universe of interest, while anything else constitutes its surroundings. These two parts may be divided by a real boundary or an imaginary conceptual one. According to the problem at hand, what constitutes the system and what constitutes its surroundings needs to be clearly stated to remove any ambiguity. It is useful to classify systems with respect to the exchanges that may take place in three different categories.

	Exchange with the surroundings		
	Matter	Heat	Work
Isolated	No	No	No
Closed	No	Yes (a) / No (b)	Yes
Open	Yes	Yes	Yes

(a) Systems with diathermal walls.
(b) Systems with adiabatic walls.

Table 1.1   Possible exchanges between a system and its surroundings.

1.2 Equilibrium

A state of equilibrium of a system is characterized by :
- Mechanical equilibrium
- Uniform pressure throughout the system
- Thermal equilibrium throughout the system
- Additionally for chemical systems : stationary concentrations for homogeneous systems or heterogeneous systems.

1.3 Thermal Reservoir. Heat Source

We often consider a **thermal reservoir**. It is a heat reservoir always at a temperature T that exchanges energy with its surroundings *exclusively* as heat. It can be approximated by a large heat capacity body or a system with phase equilibrium under a constant pressure.
We also may discuss **heat sources**. They are heat reservoirs of finite heat capacity.

1.4 Diathermal and Adiabatic Enclosures

A diathermal enclosure allows heat exchanges to take place between a system and its surroundings. An adiabatic enclosure prevents heat exchanges between a system and its surroundings.

1.5 Zeroth Law of Thermodynamics

This law derives from empirical observations and is accepted as a fundamental postulate of thermodynamics. It can be formulated as follows :

- Two systems placed in contact via a diathermal wall will, given enough time, get in thermal equilibrium.

- If one system is in thermal equilibrium with two other systems, these two systems are also in thermal equilibrium with one another.

When systems are not in thermal equilibrium, heat transfer spontaneously takes place from the hottest to the coldest one. When systems are in thermal equilibrium, heat transfer no longer takes place. Using some appropriate thermometer, the temperatures of systems in thermal equilibrium are found to be identical.

1.6 Intensive and Extensive Variables. State Functions and State Variables

1.6.1 Definitions

The thermodynamics variables mentioned here will be defined in due course.

- An *Extensive variable* depends on the size of the system.

Examples of extensive variables are U, internal energy, H, enthalpy, C_P, heat capacity at constant pressure, C_V, heat capacity at constant volume, S, entropy, A, Helmholtz energy, G, Gibbs energy, V, volume... For a system consisting of several parts, an extensive property of the ensemble of the parts is the sum of the corresponding extensive property of each of the parts. Extensive properties of a system containing a pure species are proportional to the number of moles of the species present.

- An *Intensive variable* has a uniform value in different subdivisions of a system.

Examples of intensive variables are p, pressure, T, temperature, identical in all points of the system. Molar variables or partial

1 Thermodynamic Systems : Definitions

molar variables, specific mass, mole fractions, molar heat capacity at constant pressure, $C_{p,\,m}$, have the same values in all points of one phase of the system. They may differ from one phase[†] to another.

The number of intensive variables needed to characterize the state of a system depends upon its nature. See the **phase rule** (chapters 8 and 10).
For a system with a single chemical species present as a single phase only two intensive variables need to be known. The variables we just mentioned are called **state variables** or also **state functions**. They depend only on the state of the system and not on the way the state was reached.

1.6.2   Fundamental and Auxiliary State Variables or Functions

Examples of fundamental state variables or functions are U, internal energy, V, volume, S, entropy, p, pressure, T, temperature. Auxiliary state variables are obtained starting from the fundamental ones. Examples of auxiliary state functions are H, the enthalpy, A, the Helmholtz energy, C_V, the heat capacity at constant volume, G, the Gibbs energy. Sometimes, several names may be used for the same function. See their properties in chapters 5 and 6.

$$\left.\begin{array}{ll} H = U + pV & \text{Enthalpy} \\[4pt] A = U - TS & \begin{array}{l}\text{Helmholtz energy}\\ \text{Helmholtz function}\\ \text{Free energy}\end{array} \\[4pt] G = H - TS = U + pV - TS & \begin{array}{l}\text{Gibbs energy}\\ \text{Gibbs function}\\ \text{Free enthalpy}\end{array} \end{array}\right\} \quad (1.1)$$

We will see that some of the variables are obtained by differentiating other variables :

$$\left.\begin{array}{ll} \left(\dfrac{\partial U}{\partial T}\right)_V = C_V & \text{heat capacity at constant volume} \\[8pt] \left(\dfrac{\partial H}{\partial T}\right)_p = C_p & \text{heat capacity at constant pressure} \end{array} \quad \begin{array}{l} -\left(\dfrac{\partial U}{\partial V}\right)_{S,\,n_i} = p \quad \text{pressure} \\[8pt] \left(\dfrac{\partial U}{\partial S}\right)_{V,\,n_i} = T \quad \text{temperature} \end{array}\right\} \quad (1.2)$$

[†] See chapter 8 for the definition of a phase.

1.6.3 Thermal Coefficients

Thermal coefficients indicate how volume or pressure is affected by a change in temperature. For a pure compound in a single phase, we have :

$$V = V(p, T, n) \quad (1.3)$$

$$dV = \left(\frac{\partial V}{\partial T}\right)_{p,n} dT + \left(\frac{\partial V}{\partial p}\right)_{T,n} dp + \left(\frac{\partial V}{\partial n}\right)_{p,T} dn \quad (1.4)$$

Isobaric coefficient of thermal expansion, or *isobaric expansivity*

$$\alpha = \frac{1}{V}\left(\frac{\partial V}{\partial T}\right)_{p,n} \quad (1.5)$$

Isothermal compressibility coefficient

$$\kappa = -\frac{1}{V}\left(\frac{\partial V}{\partial p}\right)_{T,n} \quad (1.6)$$

For a closed system, $dn = 0$.

$$\left.\begin{array}{l} dV = \alpha V dT - \kappa V dp = V(\alpha\, dT - \kappa\, dp) \\[6pt] d\ln V = \alpha\, dT - \kappa\, dp \end{array}\right\} \quad (1.7)$$

Constant volume thermal expansion coefficient

$$\beta = \frac{1}{p}\left(\frac{\partial p}{\partial T}\right)_{V,n} \quad (1.8)$$

Example

Assume that α and κ are constant. By integration of the second Eq. of 1.7, we have :

$$\left.\begin{array}{l} \ln\dfrac{V}{V_0} = \alpha(T - T_0) - \kappa(p - p_0) \\[4pt] \Downarrow \\[4pt] V = V_0 e^{\alpha(T - T_0) - \kappa(p - p_0)} \end{array}\right\}$$

Using as numerical values $\alpha = 2\,10^{-4}\,K^{-1}$ and $\kappa = 4\,10^{-11}\,Pa^{-1}$, let us find the volume, at 10^8 Pa and 800 K, of a liquid sample of 1 cm³ at 10^5 Pa and 300 K. We have :

$$V = 10^{-6} e^{2\,10^{-4}\cdot(800-300)\,-\,4\,10^{-11}\cdot(10^8-10^5)} \simeq 1.047\,10^{-6}\,m^3 = 1.047\,cm^3$$

The increase in temperature increases the volume of the sample, while the pressure increase, decreases it.

1 Thermodynamic Systems : Definitions

1.7 Change of a State Variable as the Result of a Thermodynamic Process

1.7.1 General Process

For a state variable, X, $(X_F - X_I)$ is independent of the path[†] used for the process. The intermediate states of the system are irrelevant.

$$(X_F - X_I)_{path\ 1} = (X_F - X_I)_{path\ 2} \tag{1.9}$$

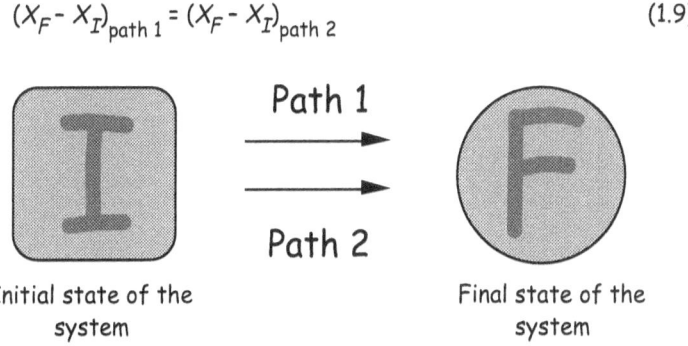

Figure 1.2 Change of a system via two different paths.

1.7.2 Cyclic Process

Consider a thermodynamic change of a system to some intermediate state via path 1. Then along path 2, bring the system back to its initial state. This process is a *cyclic process* (illustrated in figure 1.3). The change of X is zero for a cyclic process.

$$\oint dX = \int_I^{Int.} dX + \int_{Int.}^I dX = 0 \tag{1.10}$$

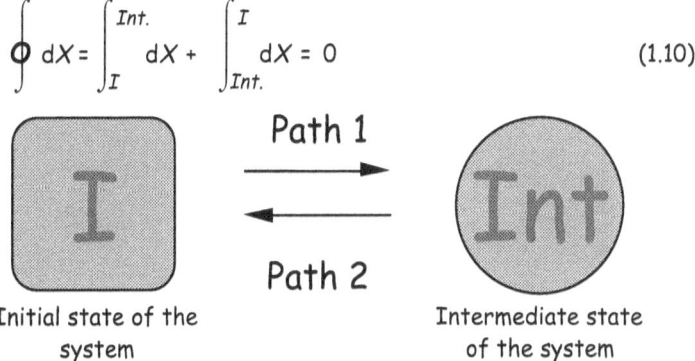

Figure 1.3 Schematic representation of a cyclic process.

[†] The path corresponds to a certain way of carrying out the process.

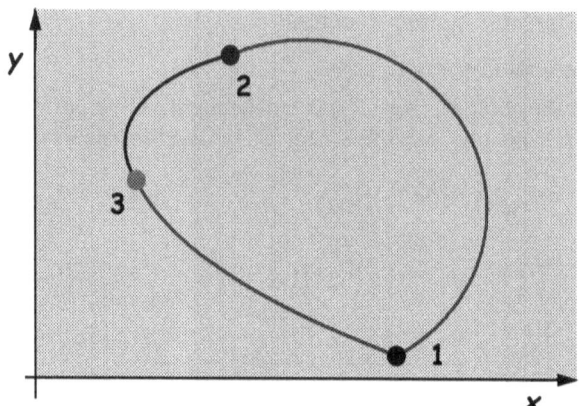

Figure 1.4 Example of a cyclic process. The variables that define the states in this case are x and y.

Example of a cyclic process: the initial state (1) and final state as displayed in Fig. 1.4 are identical.

$$\left. \oint dV = \int_1^2 dV + \int_2^3 dV + \int_3^1 dV \\ = (V_2 - V_1) + (V_3 - V_2) + (V_1 - V_3) = 0 \right\} \quad (1.11)$$

There is no volume change. The change of **any state variable** is zero for any cyclic process.

A variable X is a state variable (or state function) if its change for a cyclic process is zero. X is also a state variable if its change for a general process depends only on the initial and final states of the system and not on the way the change is achieved. The differential form dX is then called an **exact differential**. The line integral of an exact differential is independent of the path of integration.

1.7.3 Mathematical Characteristics of a State Function

Consider the differential expression:

$$dX = A\,dx + B\,dy + C\,dz \qquad (1.12)$$

In this instance, X depends on the three variables of state x, y, z.

$$dX = \left(\frac{\partial X}{\partial x}\right)_{y,z} dx + \left(\frac{\partial X}{\partial y}\right)_{z,x} dy + \left(\frac{\partial X}{\partial z}\right)_{x,y} dz \qquad (1.13)$$

1 Thermodynamic Systems : Definitions

The integral of dX is X. We have :

$$A = \left(\frac{\partial X}{\partial x}\right)_{y,z} \quad B = \left(\frac{\partial X}{\partial y}\right)_{z,x} \quad C = \left(\frac{\partial X}{\partial z}\right)_{x,y} \tag{1.14}$$

According to *Schwarz theorem*, the second derivative of a state function X with respect to two variables is independent of the order in which the derivatives are calculated. The differential expression is *exact* if :

$$\left.\begin{array}{l}\left(\dfrac{\partial A}{\partial y}\right)_{z,x} = \dfrac{\partial^2 X}{\partial y\,\partial x} = \dfrac{\partial^2 X}{\partial x\,\partial y} = \left(\dfrac{\partial B}{\partial x}\right)_{y,z} \\[6pt] \left(\dfrac{\partial B}{\partial z}\right)_{y,x} = \dfrac{\partial^2 X}{\partial z\,\partial y} = \dfrac{\partial^2 X}{\partial y\,\partial z} = \left(\dfrac{\partial C}{\partial y}\right)_{x,z} \\[6pt] \left(\dfrac{\partial C}{\partial x}\right)_{y,z} = \dfrac{\partial^2 X}{\partial x\,\partial z} = \dfrac{\partial^2 X}{\partial z\,\partial x} = \left(\dfrac{\partial A}{\partial z}\right)_{x,y}\end{array}\right\} \Rightarrow \left\{\begin{array}{l}\left(\dfrac{\partial A}{\partial y}\right)_{z,x} = \left(\dfrac{\partial B}{\partial x}\right)_{y,z} \\[6pt] \left(\dfrac{\partial B}{\partial z}\right)_{y,x} = \left(\dfrac{\partial C}{\partial y}\right)_{x,z} \\[6pt] \left(\dfrac{\partial C}{\partial x}\right)_{y,z} = \left(\dfrac{\partial A}{\partial z}\right)_{x,y}\end{array}\right\}$$

(1.15)

Example
Consider the function $F(x,y,z) = x^3 + 5x^2y - 2z^3x$. The differential dF is :
$dF = (3x^2 + 10xy - 2z^3)\,dx + 5x^2\,dy - 6z^2\,x\,dz$
The following second order differentials are equal :
$\left(\dfrac{\partial^2 F}{\partial x\,\partial y}\right) = \dfrac{\partial}{\partial x}(5x^2) = 10x = \left(\dfrac{\partial^2 F}{\partial y\,\partial x}\right) = \dfrac{\partial}{\partial y}(3x^2 + 10xy - 2z^3)$

1.8 Reversible and Irreversible Processes

A *reversible process* is carried out via a *continuous sequence of equilibrium states*.
- The intensive variables (p, T, chemical potentials) have a uniform value throughout the system.
- If and when the system is in contact with a thermal reservoir, the system temperature is equal to the temperature of the thermal reservoir.
- Losses are negligible (no friction, no viscosity).
- The forces applied to movable parts of the system are zero.

All other processes that can take place (any real processes) are *irreversible*.

1.9 Equation of State

An *equation of state* is a relationship between some of the variables of state of a pure species (usually a fluid) that completely defines the state of the system. For a pure gas: pressure, p, molar volume, V_m, and thermodynamic temperature, T. As an example, we have for a pure ideal gas:

$$p\,V_m = R\,T \qquad \text{where} \qquad V_m = \frac{V}{n} \qquad (1.16)$$

R = *gas constant* (R = 8.314472 J K^{-1} mol^{-1}), n is the number of moles of gas. The pressure is in Pascal (Pa), the volume in cubic meters (m^3), the temperature in kelvin (K).

2. Work

2.1 Sign Convention for Energy Exchange

When a thermodynamic system *receives energy* in any form, it is counted as a positive quantity. When a system *gives up energy* to some part of its surroundings, it is counted as a negative quantity. Thermodynamic laws are expressed in such a way that the sign of the energy exchanges are always accounted for from the system point of view. This sign convention is now universally adopted for its simplicity and ease of use. It is good practice to always consider algebraic thermodynamic quantities.

2.2 Mechanical Work

2.2.1 Definition

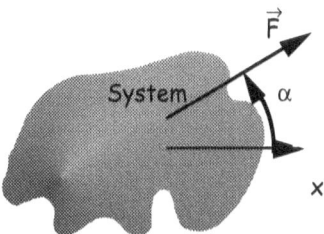

Figure 2.1 System submitted to an external force.

Consider a system submitted to an *external force*, $\vec{F}$. When the application point of the force moves by $d\vec{x}$, the algebraic mechanical work done on the system is given by:

$$\left. \begin{array}{l} dw = \vec{F} \cdot d\vec{x} \\ dw = |\vec{F}| \cos \alpha \, dx \\ w = \displaystyle\int_{x_1}^{x_2} |\vec{F}| \cos \alpha \, dx \\ \text{for a finite displacement from } x_1 \text{ to } x_2 \text{ along the } x \text{ axis} \end{array} \right\} \quad (2.1)$$

The elementary work dw expressed in Eq. 2.1 has the appropriate sign with respect to the sign convention of § 2.1.

If the external force vector $\vec{F}$ has a component in the same direction as the displacement taking place, then the corresponding elementary work is positive.

Example
If the application point of a force of 10 N (the unit of force is the newton) is displaced 1 m in its direction (angle $\alpha = 0$), the work done, on the system it applies to, is :
$$w = 10 \cdot 1 = 10 \text{ J}$$
Energy is measured in joules, symbol J.

2.2.2 Work and Volume Change

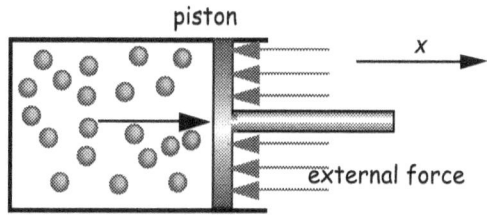

Figure 2.2 Gas inside a cylinder closed by a piston. The external force may be caused by a pressure.

The system is a gas in a cylinder closed by a piston of area A. Everything else is called the *surroundings* of the system. If V increases during the process, the gas is said to undergo an *expansion*. If V decreases during the process, the gas is said to undergo a *compression*. The external force exerted on the piston is due to the external pressure (Fig. 2.2).
If $p_{int} > p_{ext}$, the piston has a spontaneous tendency to move to the right, opposite to the external force. The magnitude of the external force on the piston is $p_{ext} A$ towards the left.
For $dx > 0$, the system does work on the surroundings, dw is negative.

$|\vec{F}| = p_{ext} A$
opposite to the direction of the displacement

$dw = -p_{ext} A\, dx = -p_{ext}\, dV$ Expression always valid

$dV = A\, dx$
volume increase during the process
$dx > 0 \Rightarrow dV > 0$ and $dw < 0$

(2.2)

A displacement in the direction of the external force occurs when $p_{int} < p_{ext}$, then $dx < 0$ and $dw > 0$. The expression for the (algebraic) work done by the system due to its change of volume is always valid.

2 Work

2.2.3 Process at Constant External Pressure

The work done on the system during the process is:

$$w = - \int_{V_I}^{V_F} p_{ext} \, dV = - p_{ext} \int_{V_I}^{V_F} dV = - p_{ext} (V_F - V_I) \qquad (2.3)$$

The work done on the system *depends on the external pressure*, i.e. the outside, and does not depend on the inside pressure of the system. If $p_{ext} < p_{int}$ and p_{ext} constant, we find:

$V_F > V_I \Rightarrow \qquad w < 0 \qquad$ expansion

If $p_{ext} = 0$ then $\quad w = 0$

Example

We can calculate the volume work done on a system during a chemical reaction that takes place at constant external pressure. Assume the following reaction is complete at 25 °C (298.15 K) and 10^5 Pa using an appropriate catalyst.

$$CO_2 + 4 H_2 \longrightarrow CH_4 + 2 H_2O \text{ (l)}$$

Starting with 0.5 mole of gas in stoichiometric proportion in the initial state, there is only 0.1 mole of gas when the reaction is complete (the water is liquid, its volume is negligible compared to the gas volume). Assuming that the gases obey the ideal gas law, the initial and final volumes of the system are:

$$V_I = \frac{n_I R T}{p} = \frac{0.5 \cdot 8.3145 \cdot 298.15}{10^5} \simeq 12.39 \cdot 10^{-3} \, m^3 = 12.39 \, l$$

$$V_F = \frac{n_F R T}{p} = \frac{0.1 \cdot 8.3145 \cdot 298.15}{10^5} \simeq 2.48 \cdot 10^{-3} \, m^3 = 2.48 \, l$$

The work done on the system is:

$$w = - p_{ext} (V_F - V_I) \simeq - 10^5 (2.48 \cdot 10^{-3} - 12.39 \cdot 10^{-3}) \simeq 991 \, J$$

The surroundings do work on the system.

2.2.4 Work during an Isothermal (Reversible) Change of an Ideal Gas

During an *isothermal reversible change*, the temperature of the system stays *constant and uniform for the entire process* which takes place as a *continuous succession of equilibrium states*.
The gas volume varies from V_I to V_F. The external force applied to the piston in every intermediate state exactly compensates the force due to the internal pressure of the system. We apply at every instant an external pressure, $p_{ext} = p_{int}$. For an infinitesimal process, the work done on the system is given by:

$$dw = - p_{ext} \, dV = - p_{int} \, dV \qquad (2.4)$$

For a *finite isothermal (reversible) process of an ideal gas*, the ideal gas law can be used since, at every moment, the gas is at equilibrium. The expression for the work done on the gas is :

$$w = -\int_{V_I}^{V_F} \frac{nRT}{V} dV = -nRT \int_{V_I}^{V_F} \frac{dV}{V} = -nRT \ln\left(\frac{V_F}{V_I}\right) \qquad (2.5)$$

Example
We can calculate the volume work done on one mole of an ideal gas when its volume doubles during an isothermal change of volume at 500 K.
$w = -1 \cdot 8.3145 \cdot 500 \cdot \ln 2 \simeq -2881 \text{ J} \simeq -2.88 \text{ kJ}$
Since w is negative, the gas does work on the surroundings.

2.3 Remarks

- For systems interacting with a force field, the gravity field for example, it is necessary to perform a global balance of the *external forces* acting on the system.

- Consider a *reversible process*. A reversal of the displacement direction, keeping all external forces unchanged, implies a change of the sign of the work done on the system. For a finite process, a change in the sign of each elementary displacement, while the system goes through the same intermediate states (identical forces), defines a new process often referred to as the **reverse process**. The work done on the system during the *reverse process* is therefore *opposite* to that done on it during the *direct process*. The change in the internal energy of the system in the reverse process is opposite to the change in the direct process, which finally implies (See Chapter 3, The First Law) also that the heat received by the system is opposite to that received in the direct process.

2.4 Electrical Work

A battery can transform potential chemical energy into electrical work. The chemical reaction is carried out at electrodes via redox processes. The **source voltage** E (electromotive force – e.m.f.) is defined to always be positive.

$$E = (\phi_+ - \phi_-) \qquad (2.6)$$

E does not depend on the reference for the potentials ϕ_+ and ϕ_-. The work done on the battery when a charge dQ goes from a potential ϕ_+ to a potential ϕ_- ($\phi_+ > \phi_-$) is:

$$dw = -(\phi_+ - \phi_-)dQ = -EdQ \qquad (2.7)$$

For dQ positive, dw is negative, work is done by the battery on the outside. This corresponds to the normal operation of a battery.
In the charging process, dQ is negative. Energy is stored in the battery in a chemical form. The electric potential (the voltage) has units of Volts (V) and the charge has units of Coulombs (C).
The charge of one mole of a singly charged ion is 1 Faraday, $F = N_A e = 96485.3415 \, C \, mol^{-1}$, N_A is Avogadro's constant, the number of particles present in one mole of a species, $N_A = 6.02214199 \, 10^{23} \, mol^{-1}$.

2.5 Various Forms of Energy

Energy can be found in various forms such as electrical, mechanical and thermal energy. Other forms have their origin in chemical, electromagnetic, gravitational and nuclear interactions.

2.6 Various Expressions for Work

When work has origins other that those mentioned until now, it is necessary to use appropriate expressions for work in each case.
If the area A of a system, with interfacial tension σ, increases by dA, the work done on the system is:

$$dw = \sigma dA \qquad (2.8)$$

For an increase dz of the altitude of an object of mass m in the gravity field g, the work done on the system is:

$$dw = mg\,dz \qquad (2.9)$$

Lengthening an elastic material by dl under tension $\mathscr{T}$ does work on the system, it is given by:

$$dw = \mathscr{T}dl \qquad (2.10)$$

The work done on a substance in a magnetic field B with a magnetic moment dM is:

$$dw = B\,dM \qquad (2.11)$$

3. First Law of Thermodynamics

3.1 Introduction

Energy exchanges between bodies do occur. One can observe :
- Energy exchange as either work or heat
- Conversion of work to heat or conversely of heat to work
- Energy exchange simultaneously as both work and heat.

The *First Law* formulates some of the rules applicable to energy exchanges. It can be simply stated as : *energy is conserved*.

3.2 The Joule Experiment

By using the fall of a weight in the gravity field of the earth, a paddle wheel that turns in liquid water causes a rise in its temperature. Joule showed that the same temperature rise could be obtained using an electrical resistor heated by an electric current.

Work can be transformed into heat. Heat and work are of the same nature and constitute different forms of energy. They are expressed in the same units.

The joule (symbol J) is the SI unit of energy. One still encounters in practical applications the calorie (symbol cal) to express amounts of heat. One calorie corresponds to the amount of heat that is needed to get one gram of water from $14.5°C$ to $15.5°C$. We have the equivalence 1 cal = 4.1840 J.

3.3 Internal Energy. First Law

3.3.1 General Aspects. Expression of the First Law

The *internal energy* U depends on the state of the system and accounts for all energy exchanges.

We ignore the potential energy of the system in the gravity field of the earth or in external fields.

> The internal energy of a system is a state function. The differential of the internal energy U is an exact differential.

3.3.2 Closed System - Adiabatic Process

A *closed system* contained inside an adiabatic enclosure does not exchange heat. When work is done on such a system, it changes its internal energy.

$$q = 0 \qquad U_F - U_I = w_{adiabatic} \tag{3.1}$$

Example
The work done on a gas during an adiabatic process is 1000 J. The change of its internal energy for this process is :
$$U_F - U_I = w_{adiabatic} = 1000 \text{ J}$$

3.3.3 Closed System - General Process. First Law

Consider a *closed system* exchanging energy with its surroundings as both *work* and *heat*.
Between two given states of the system, the change in the internal energy of the system is independent of the process. This statement actually constitutes the *first law*.

$$U_F - U_I = w + q \tag{3.2}$$

The work done on the system and the heat received by the system depend on the process.

Example
As a result of a thermodynamic process, 1000 J of work is done on a system in contact with a thermal reservoir at 300 K. The system receives − 700 J of heat (The system actually transfers heat to the thermal reservoir). The change in the internal energy of the system is :
$$U_F - U_I = w + q = 1000 - 700 = 300 \text{ J}$$

3.3.4 Closed System - Cyclic Process

For a cyclic process, the initial and final states are identical :

$$U_F = U_I \quad \Rightarrow \quad \Delta U = 0 = w + q \tag{3.3}$$

The sum of the work and heat received by a system during a cyclic process is zero.

3.3.5 Infinitesimal Process

For an infinitesimal process, we can write Eq. 3.2 in a differential form :

$$dU = dw + dq \tag{3.4}$$

dU is an exact differential while, in general, dq and dw are not.

4. Second Law of Thermodynamics

4.1 Kelvin Formulation of the Second Law

The second law of thermodynamics can be formulated in the following way:
> Using a system which undergoes a cyclic thermodynamic process, it is impossible to obtain useful work if, globally, heat is only exchanged with one thermal reservoir.

A process where heat is only exchanged with a single thermal reservoir is called a **monothermal process**. Useful work to us corresponds to $w < 0$. In a mathematical form, we can write the Kelvin formulation of the second law as:

$$w \geq 0 \quad \text{cyclic process - single thermal reservoir affected} \quad (4.1)$$

As a consequence, for a *cyclic monothermal* process, we have:

$$\Delta U = w + q = 0 \quad \Rightarrow \quad q \leq 0 \quad (4.2)$$

It can be shown that for a *cyclic monothermal reversible* process $w = 0$.

4.2 Carnot Cycle. Heat Engine. Thermodynamic Temperature

The *Carnot cycle* is a *dithermal reversible cycle*. The system exchanges heat *reversibly* with two thermal reservoirs at temperatures, T_ℓ (cold) and T_h (hot) (see Fig. 4.1).

Due to the reversibility requirements:

- The system can only be brought in contact with a thermal reservoir when its temperature is identical to the temperature of the thermal reservoir. While in contact with the thermal reservoir, the process is *isothermal*, the system temperature is uniform.

- When the system is not in contact with one of the thermal reservoirs, the process that it undergoes *can only be adiabatic*. The system then receives or provides work. Its internal energy changes and so does its temperature.

Therefore, the *Carnot cycle* comprises the following steps:

- 1-2 Isothermal (reversible) process in contact with the thermal reservoir at high temperature T_h, the system receives q_h

- 2-3 Reversible adiabatic process that changes the system temperature from T_h to T_ℓ
- 3-4 Isothermal (reversible) process in contact with the thermal reservoir at low temperature T_ℓ, the system receives q_ℓ
- 4-1 Reversible adiabatic process that changes the system temperature from T_ℓ to T_h.

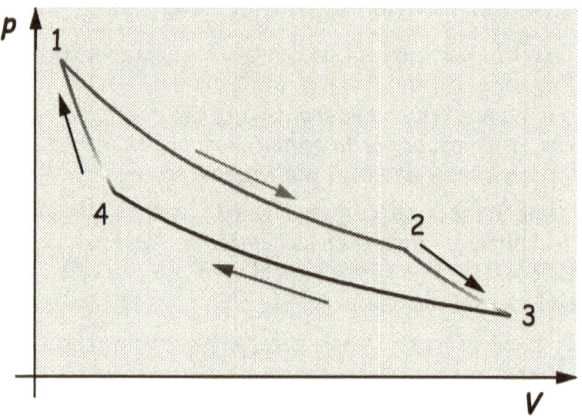

Figure 4.1 Example of a Carnot engine cycle.

An engine provides work to a user (w is negative). For a **Carnot engine cycle** (or **forward Carnot cycle**), it can be shown that we have the following relations:

$$w + q_h + q_\ell = 0 \atop w < 0 \quad q_\ell < 0 \quad \Rightarrow \quad \left\{ {q_h > 0 \atop \text{and} \quad q_h > |q_\ell|} \right\} \qquad (4.3)$$

where w is the algebraic work done on the system during the cycle.

For any system undergoing a *reversible Carnot cycle*, one can show that the ratio of *the amount of heat received by a system* from each thermal reservoir is independent of the system and depends only on the thermal reservoirs.

This fact is used to define a temperature scale that has a universal character, the **thermodynamic temperature**. We write:

$$\frac{T_h}{T_\ell} = -\frac{q_h}{q_\ell} \quad \Rightarrow \quad \frac{q_h}{T_h} + \frac{q_\ell}{T_\ell} = 0 \qquad (4.4)$$

This choice implies that the temperatures of the thermal reservoirs have the same sign on this scale and are always positive. We also have:

$$T_h > T_\ell \qquad (4.5)$$

4 Second Law of Thermodynamics

The temperature of the hot thermal reservoir is larger than that of the cold thermal reservoir.

Unique temperature values are obtained by selecting as a reference for the thermodynamic temperature the triple point of water, temperature at which ice, liquid water and water vapor are at equilibrium[†]. The vapor pressure of liquid water is then 6.11 mbar (hPa). The numerical value selected for that temperature is $T_{tp\,H_2O}$ = 273.16 K, the temperature unit is the kelvin (symbol K). With this choice, the temperature at which ice melts under a pressure of 1 bar is 273.15 K and the boiling temperature of water under the normal pressure of 1 atm is 373.15 K. This selection leads, for these two phenomena, to a difference of 100 K on the thermodynamic scale and a 100°C difference on the Celsius (centigrade) temperature scale.

> All the equations of thermodynamics are only valid if we use the thermodynamic temperature scale.

Under the SI standard pressure of 1 bar, the boiling temperature of water is 372.8 K.

Example

As a result of a Carnot cycle, a system provides 1000 J of work. Therefore – 1000 J of work is done on a system that comes in contact with thermal reservoirs at 1500 K and 300 K. This system actually does work on its surroundings. The ratio of the amount of heat received by the system from the thermal reservoirs is :

$$\frac{T_h}{T_\ell} = -\frac{q_h}{q_\ell} \Rightarrow \frac{q_h}{q_\ell} = -\frac{1500}{300} = -5$$

The amounts of heat received by the thermal reservoirs are :

$$w + q_h + q_\ell = 0 \Rightarrow -1000 - 5q_\ell + q_\ell = 0 \Rightarrow \begin{array}{l} q_\ell = -250\text{ J} \\ q_h = 1250\text{ J} \end{array}$$

We find that 250 J of energy is "wasted" to obtain 1000 J of work.

4.3 Entropy. Reversible and Irreversible Processes. Equilibrium

4.3.1 Definition

We define a new function called entropy. During an infinitesimal *reversible* change of a system in contact with a thermal reservoir at temperature T, **the entropy of the system** varies by :

$$dS_{sys} = \frac{dq_{rev}}{T} \tag{4.6}$$

[†] Phase equilibrium is presented in chapter 8.

dq_{rev} is the amount of heat received by the system during the reversible process. *The entropy S of a system is a state function.* During an *adiabatic **reversible** process*, the system entropy does not vary since dq_{rev} = 0. An *adiabatic reversible process* is also called an ***isentropic process***. Equation 4.4 indicates that as a result of a Carnot cycle, the entropy of the system does not change.

Example
During a reversible adiabatic process, 2500 J of work is done on a system. The internal energy change of the system is :

$U_F - U_I = w_{adiabatic}$ = 2500 J

The entropy change of the system is :

$S_F - S_I$ = 0 J K^{-1}

4.3.2 The Second Law, Entropy and Spontaneity of Processes

If a closed adiabatic system evolves irreversibly and thus spontaneously, its entropy in its final state is larger than its entropy in its initial state. This statement can be shown to be a direct consequence of the Kelvin formulation given at the beginning of this chapter[†].

$$S_{final} > S_{initial} \quad \Leftrightarrow \quad S_{final} - S_{initial} > 0 \quad (4.7)$$

Note that any spontaneous process is irreversible. It usually takes place as a consequence of some change imposed to the system or to part of it. Since reversible adiabatic processes are isentropic, we can also state :

It is impossible to imagine a process for a closed adiabatic system that would result in a decrease of its entropy.

Consider now a system that can be in contact with a thermal reservoir at temperature T_{therm}. The *global system* (system of interest plus the thermal reservoir) is an adiabatic system. For an irreversible process, *the entropy change of the global system is positive.*
For any process of a *closed adiabatic global system*, we have the relation :

global system = system of interest plus thermal reservoir

$$dS_{global} \geq 0 \quad (S_F - S_I)_{global} \geq 0 \quad (4.8)$$

The equality applies to a *reversible process* and the inequality to *an irreversible one*. It is often possible to consider that a system and its surroundings constitute an adiabatic closed system. The entropy of an adiabatic closed system does not change during a *reversible process*.

[†] The Bases of Chemical Thermodynamics, M. Graetzel & P. Infelta, chapter 4.

Example

Consider a system undergoing a Carnot cycle. For this cyclic process the entropy change of the system is zero since the initial and final states of the system are the same. The entropy changes of the thermal reservoirs are:

$$(S_F - S_I)_{\text{therm } h} = -\frac{q_h}{T_h} \qquad (S_F - S_I)_{\text{therm } \ell} = -\frac{q_\ell}{T_\ell}$$

In virtue of Eq. 4.4, the global entropy change (system plus thermal reservoirs) is found to be zero.

Example

Consider two containers with adiabatic walls separated by a stopcock. Initially, one container contains one mole of an ideal gas at a pressure of 1 bar, while container 2 is evacuated. Both containers have the same volume. The initial temperature of the gas is 25°C.

We open the stopcock. An irreversible expansion takes place and the gas occupies the entire volume at its disposal. Since the system is adiabatic, $q = 0$, and the system volume does not change, $w = 0$, the internal energy of the gas does not change.

$$U_F - U_I = 0 = w + q$$

The internal energy of an ideal gas depends only on its temperature (chapter 7). When the system has reached equilibrium, its temperature has not changed. The change in the entropy of the gas can be obtained starting from (Eq. 4.13) using also the ideal gas law:

$$dU = 0 = -p\,dV + T\,dS \quad \Rightarrow \quad dS = \frac{p}{T}dV = nR\frac{dV}{V}$$

$$S_F - S_I = nR \int_{V_I}^{V_F} \frac{dV}{V} = nR \ln\frac{V_F}{V_I} = 8.3145 \cdot \ln 2 \approx 5.76 \text{ J K}^{-1}$$

For this irreversible adiabatic process, the entropy of the system increases.

4.3.3 System in Contact with a Single Thermal Reservoir

Consider a process of a *closed system* that can exchange heat with only one thermal reservoir at temperature T_{therm} during part of the process. The global system (subscript $_{\text{global}}$) made of the system and the thermal reservoir, is a closed adiabatic system. The system receives q_{sys}. The amount of heat received by the thermal reservoir is:

$$q_{\text{therm}} = -q_{\text{sys}} \tag{4.9}$$

Since the global change of entropy cannot be negative, we have:

$$(S_F - S_I)_{global} = (S_F - S_I)_{sys} + (S_F - S_I)_{therm}$$

$$= (S_F - S_I)_{sys} - \frac{q_{sys}}{T_{therm}} \geq 0 \qquad (4.10)$$

From which we can write:

$$q_{sys} \leq T_{therm}(S_F - S_I)_{sys} = (q_{sys})_{rev} \qquad (4.11)$$

For a system going from state I to state F, the amount of heat the system receives is smaller for an irreversible process than for a reversible one. The last equality in Eq. 4.11 is justified by imagining a reversible process that takes the system from its initial to its final state. We carry out a reversible adiabatic process to bring the system temperature to T_{therm}, followed by a (reversible) isothermal process at T_{therm} and finally another adiabatic reversible process to bring the system to its final state, thus the expression for $(q_{sys})_{rev}$.

Example

Here, we show the reversible path for a gaseous system going from an initial state I to a final state F, via a reversible adiabatic process followed by an isothermal process at the temperature T_{therm} and finally another reversible adiabatic process to bring the system temperature to T_F. The entropy of the system varies only during the isothermal part of the process.

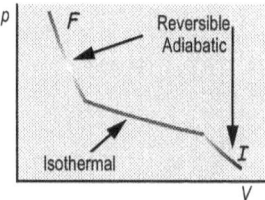

4.3.4 Clausius Inequality

Consider a system that, during a cycle, comes in contact with several thermal reservoirs at different temperatures, $T_1, T_2, ..., T_i, ..., T_n$. The entropy change of the system during the cycle is of course zero since entropy is a state function. Using the inequality established in (4.10) for each thermal reservoir and summing for all thermal reservoirs, we can write:

$$0 = \Delta S_{sys} = \sum_{i=1}^{n} \Delta S_{i\,sys} \geq \sum_{i=1}^{n} \frac{q_{i\,sys}}{T_i} \Rightarrow \sum_{i=1}^{n} \frac{q_{i\,sys}}{T_i} \leq 0 \qquad (4.12)$$

for a system in contact with two thermal reservoirs

$$\frac{q_h}{T_h} + \frac{q_\ell}{T_\ell} \leq 0$$

4 Second Law of Thermodynamics

$q_{i\,sys}$ is the amount of heat received by the system from the thermal reservoir at temperature T_i. The equality is valid for a reversible cycle while the inequality applies to any irreversible cycle.

Example
For a cyclic process of a system entering in contact with n thermal reservoirs, the entropy change of the system is zero. From the amount of heat received by the thermal reservoir at temperature T_i, we find the entropy change of that thermal reservoir to be:

$$(S_F - S_I)_{therm\,i} = -\frac{q_{i\,sys}}{T_i} \quad \Rightarrow \quad (S_F - S_I)_{global} = \sum_{i=1}^{n} -\frac{q_{i\,sys}}{T_i} \geq 0$$

We find using 4.12 that the global entropy change is positive for an irreversible cyclic process and zero for a reversible one.

4.3.5 Differential Expression for the Internal Energy and Enthalpy of a Closed System

Assume that *work due to volume change* is the only type of work done on a closed system (without chemical reactions). For a reversible process, the external pressure is equal to the internal pressure (system pressure) and the system temperature is equal to a thermal reservoir temperature if and when there is thermal contact (needed of course for heat exchange).

$$\left. \begin{array}{l} p_{ext} = p_{int} = p \\ dU = dw + dq_{rev} = -p_{ext}\,dV + T\,dS = -p\,dV + T\,dS \end{array} \right\} \quad (4.13)$$

The resulting expression is *valid for any process*, even irreversible processes, since U is a function of state and dU is an exact differential.

Enthalpy is another extensive state function. We can find the expression for the differential of enthalpy from the expression just obtained for dU:

$$H = U + pV \quad \Rightarrow \quad dH = dU + p\,dV + V\,dp = T\,dS + V\,dp \quad (4.14)$$

Example
Let us find the differential expression for the entropy change of n moles of a pure ideal gas. From the two different expressions of the differential of the internal energy of n moles of an ideal gas, we obtain:

$$dU = n\,C_{V,m}\,dT = -p\,dV + T\,dS \quad \Rightarrow \quad dS = n\frac{C_{V,m}}{T}dT + \frac{p}{T}dV = n\left(\frac{C_{V,m}}{T}dT + \frac{R}{V}dV\right)$$

$C_{V,m}$ is the molar heat capacity of the gas at constant volume (see chapters 5 and 6). This differential expression can be integrated assuming that the molar heat capacity of the gas is constant. We obtain the expression for the change in entropy of an ideal gas characterized by its temperature and volume :

$$S_F - S_I = n\left[C_{V,m}\ln\left(\frac{T_F}{T_I}\right) + R\ln\left(\frac{V_F}{V_I}\right)\right]$$

Using the ideal gas law and Eq. 5.52, we find expressions for the entropy change when the states are characterized by other variables :

$$S_F - S_I = n\left[C_{V,m}\ln\left(\frac{p_F V_F}{p_I V_I}\right) + R\ln\left(\frac{V_F}{V_I}\right)\right] = n\left[C_{V,m}\ln\left(\frac{p_F}{p_I}\right) + C_{p,m}\ln\left(\frac{V_F}{V_I}\right)\right]$$

$$S_F - S_I = n\left[C_{V,m}\ln\left(\frac{T_F}{T_I}\right) + R\ln\left(\frac{p_I T_F}{p_F T_I}\right)\right] = n\left[C_{p,m}\ln\left(\frac{T_F}{T_I}\right) - R\ln\left(\frac{p_F}{p_I}\right)\right]$$

$C_{p,m}$ is the molar heat capacity of the gas at constant pressure (chapters 5 and 6).

4.3.6 Equilibrium Condition

Let us now consider a chemical system in an adiabatic enclosure. An irreversible, thus spontaneous, process can take place only if it results in an increase in entropy of the system. Such a process keeps on going until the entropy of the system can no longer increase. The system is then at equilibrium and its entropy has reached a maximum.

> When no spontaneous process takes place, the entropy of a closed adiabatic system is maximal and the system is at equilibrium.

4.3.7 Thermal Equilibrium

Consider two identical blocks of matter in thermal contact with each other ($C_{p1} = C_{p2} = C_p =$ Heat capacities of each block at constant pressure). The system is adiabatic. The heat transfer is assumed to be very slow (then the temperature of each block stays uniform). The amount of heat received by a system during an isobaric (constant pressure) process is the change of its *enthalpy* (See § 5.2).
Since the global system receives no heat, the change of its enthalpy is zero. We obtain the final temperature, T_F:

$$\left.\begin{array}{r}H_F - H_I = \displaystyle\int_{T_1}^{T_F} C_p \, dT + \int_{T_2}^{T_F} C_p \, dT = C_p(T_F - T_1) + C_p(T_F - T_2) = 0 \\ \Rightarrow \quad T_F = \dfrac{T_1 + T_2}{2} \end{array}\right\}$$

(4.15)

The entropy change of the system during the process is :

4 Second Law of Thermodynamics

Figure 4.2 Heat exchange between two blocks of matter at different temperatures. Globally, the system receives no heat.

$$S_F - S_I = \int_{T_1}^{T_F} \frac{C_p}{T} dT + \int_{T_2}^{T_F} \frac{C_p}{T} dT = C_p \ln \frac{T_F}{T_1} + C_p \ln \frac{T_F}{T_2} \quad (4.16)$$

$$\left. \begin{array}{c} S_F - S_I = C_p \ln \dfrac{T_F^2}{T_1 T_2} = C_p \ln \dfrac{(T_1 + T_2)^2}{4 T_1 T_2} \\ \Downarrow \\ S_F - S_I = C_p \ln \left(1 + \dfrac{(T_1 - T_2)^2}{4 T_1 T_2} \right) \end{array} \right\} \quad (4.17)$$

The entropy change of this closed adiabatic system is always positive, except if both blocks are at the same temperature. Then no process takes place, since the blocks are already in thermal equilibrium.

Example

Let us examine how the ratio $(S_F - S_I)/C_p$ varies with the ratio T_2/T_1 of the initial temperatures of the blocks. We have :

$$\frac{S_F - S_I}{C_p} = \ln \left[1 + \frac{\left(1 - \dfrac{T_2}{T_1}\right)^2}{4 \dfrac{T_2}{T_1}} \right]$$

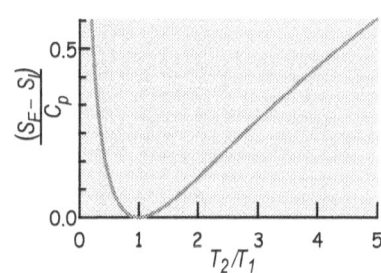

We see that if, in the initial state, the temperatures of the blocks are different, the heat transfer implies an increase of the entropy of the system. A bigger difference in the initial temperatures results in a larger entropy increase.

4.3.8 Isothermal (Reversible) Expansion of an Ideal Gas

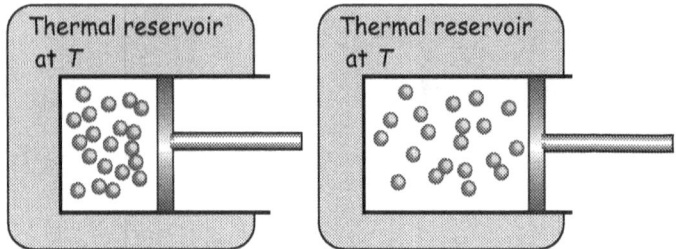

Figure 4.3 Isothermal expansion of an ideal gas. The walls of the cylinder are in thermal contact with a thermal reservoir. The mobile piston is adiabatic.

A system contains n moles of an ideal gas in a cylinder closed by a mobile adiabatic piston (Fig. 4.3). The other walls of the cylinder are in thermal contact with a thermal reservoir at temperature T.

The external pressure stays identical to the pressure inside the system, $p_{ext} \simeq p_{int}$, for all states during the process. The internal energy of the ideal gas remains constant (See chapter 7). We have:

$$w_{sys} = -\int_{V_I}^{V_F} p_{ext}\, dV = -\int_{V_I}^{V_F} \frac{nRT}{V}\, dV = -nRT \ln\left(\frac{V_F}{V_I}\right)$$

$$U_F - U_I = 0 = w_{sys} + q_{sys} \qquad q_{sys} = nRT \ln\left(\frac{V_F}{V_I}\right)$$

(4.18)

The entropy change of the system is:

$$(S_F - S_I)_{sys} = \frac{1}{T}\int_I^F dq_{rev} = \frac{q_{sys}}{T} = nR \ln\left(\frac{V_F}{V_I}\right) \qquad (4.19)$$

The entropy change of the thermal reservoir is:

$$(S_F - S_I)_{therm} = \frac{1}{T}\int_I^F dq_{therm} = \frac{-q_{sys}}{T} = -nR \ln\left(\frac{V_F}{V_I}\right) \qquad (4.20)$$

The entropy change of the global (closed adiabatic) system is zero, as expected for a reversible process.

$$(S_F - S_I)_{global} = (S_F - S_I)_{sys} + (S_F - S_I)_{therm} = 0 \qquad (4.21)$$

Example

Let us find the work done on 0.1 mole of an ideal gas during a reversible isothermal process at 400 K when its volume doubles.

$$w_{sys} = -nRT\ln\left(\frac{V_F}{V_I}\right) = -0.1 \cdot 8.3145 \cdot 400 \ln 2 \simeq -230.5 \text{ J}$$

The amount of heat received (from the thermal reservoir) by the system and its entropy change are:

$$q_{sys} = -w_{sys} \simeq 230.5 \text{ J} \qquad (S_F - S_I)_{sys} = nR\ln\left(\frac{V_F}{V_I}\right) \simeq 0.58 \text{ J K}^{-1}$$

The internal energy of the gas stays constant during the process. The work obtained comes integrally from the thermal reservoir. The entropy change of the thermal reservoir is:

$$(S_F - S_I)_{therm} = \frac{q_{therm}}{T} = -\frac{q_{sys}}{T} \simeq -0.58 \text{ J K}^{-1}$$

The global entropy change for this reversible process is zero.

4.3.9 Monothermal Irreversible Expansion of an Ideal Gas

Consider now the irreversible expansion of the ideal gas where the initial and final states of the gas are the same as in the case just examined (Fig.4.3).

The external pressure is now assumed to remain constant during the process and has the value $p_F < p_I$. Entropy is a state function, the entropy change of the system is:

$$(S_F - S_I)_{sys} = nR\ln\left(\frac{V_F}{V_I}\right) = nR\ln\left(\frac{p_I}{p_F}\right) \qquad (4.22)$$

The work done on the system is:

$$w_{sys} = -\int_{V_I}^{V_F} p_{ext}\, dV = -\int_{V_I}^{V_F} p_F\, dV = -p_F(V_F - V_I) \qquad (4.23)$$

The final temperature of the gas is the same as its initial temperature and equal to the thermal reservoir temperature. The first law provides:

$$U_F - U_I = 0 = w_{sys} + q_{sys} \Rightarrow q_{sys} = -w_{sys} = p_F(V_F - V_I) \qquad (4.24)$$

The entropy change of the thermal reservoir is:

$$q_{therm} = -q_{sys} = -p_F(V_F - V_I)$$

$$(S_F - S_I)_{therm} = \frac{q_{therm}}{T} = -\frac{p_F(V_F - V_I)}{T} = -\frac{p_F V_F}{T}\left(1 - \frac{V_I}{V_F}\right)$$

$$= -nR\left(1 - \frac{V_I}{V_F}\right) = -nR\left(1 - \frac{p_F}{p_I}\right)$$

$$(4.25)$$

Combining 4.22 and 4.25, we obtain the entropy change of the global system (gas plus thermal reservoir):

$$(S_F - S_I)_{global} = nR \left\{ \ln\left(\frac{p_I}{p_F}\right) - \left(1 - \frac{p_F}{p_I}\right) \right\} \qquad (4.26)$$

We show in Fig. 4.4 how the global entropy changes as a function of the ratio p_F/p_I. The entropy of the global system increases except if $p_F = p_I$ (when no process takes place).

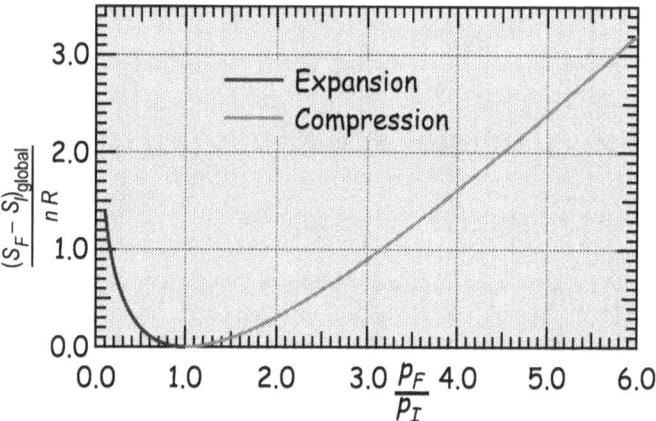

Figure 4.4 Entropy changes for a monothermal irreversible process (expansion or compression) of an ideal gas at constant external pressure p_F.

Example

Let us find the work done on 0.1 mole of an ideal gas during the irreversible monothermal process of § 4.3.9 when its volume doubles (same initial and final states as in the previous example). The temperature of the gas is 400 K in its initial and final states. The work done on the system is:

$$w_{sys} = -p_F(V_F - V_I) = -p_F V_F \left(1 - \frac{V_I}{V_F}\right)$$

$$= -nRT\left(1 - \frac{V_I}{V_F}\right) = -0.1 \cdot 8.3145 \cdot 400 \cdot 0.5 \simeq -166.3 \text{ J}$$

The amount of heat received by the system is:

$$q_{sys} = -w_{sys} \simeq 166.3 \text{ J}$$

The entropy change of the system is the same as in the reversible process (previous example) since entropy is a state function. The entropy change of the thermal reservoir is:

$$(S_F - S_I)_{therm} = \frac{q_{therm}}{T} = -\frac{q_{sys}}{T} \simeq -\frac{166.3}{400} \simeq -0.42 \text{ J K}^{-1}$$

The change of the entropy of the global system (system + thermal reservoir) is:
$$(S_F - S_I)_{global} = (S_F - S_I)_{sys} + (S_F - S_I)_{therm} \simeq 0.58 - 0.42 = 0.16 \ J\,K^{-1}$$
As expected from the second law, it is a positive quantity.

4.3.10 Reversible Adiabatic Process of an Ideal Gas

Consider n moles of an ideal gas undergoing an adiabatic reversible process. The first law implies:

$$dq = dU - dw = C_V dT + p \, dV = 0 \tag{4.27}$$

Since the process is reversible, we can express p with the ideal gas law, we obtain:

$$C_V \, dT + nRT \frac{dV}{V} = 0 \quad \Rightarrow \quad C_V dT + (C_p - C_V) T \frac{dV}{V} = 0 \tag{4.28}$$

We replaced nR by its value in terms of the heat capacities of the system (Eq. 5.52). We can divide Eq. 4.28 by the number of moles n and, assuming constant **molar heat capacities**, we can integrate the resulting differential equation.

$$C_{V,m} \ln\left(\frac{T_2}{T_1}\right) + (C_{p,m} - C_{V,m}) \ln\left(\frac{V_2}{V_1}\right) = 0 \tag{4.29}$$

$C_{p,m}$ and $C_{V,m}$ are the **molar heat capacities**. We can obtain various forms of this relation:

$$\left. \frac{T_2}{T_1} = \left(\frac{V_1}{V_2}\right)^{\left\{\frac{C_{p,m}}{C_{V,m}} - 1\right\}} = \left(\frac{V_1}{V_2}\right)^{(\gamma-1)} \quad \text{with } \gamma = \frac{C_{p,m}}{C_{V,m}} \right\} \tag{4.30}$$

$$\text{with } pV = nRT \Rightarrow p_1 V_1^{\gamma} = p_2 V_2^{\gamma}$$

The entropy change of the system is zero. This process is an **isentropic process**. During an isothermal change of an ideal gas, the product pV remains constant, while during an isentropic change, it is the product pV^{γ} that remains constant. In a diagram giving p vs. V, and at a specific point p_0, V_0, the slope of an isothermal curve is $-p_0/V_0$ while that, on an adiabatic curve going through the same point, is $-\gamma p_0/V_0$. The ratio of the heat capacities at constant pressure and at constant volume, γ, is always larger than 1 (see Fig. 4.5). The absolute value of the slope of the adiabatic curve is always larger than that of the isothermal curve going through the same point.

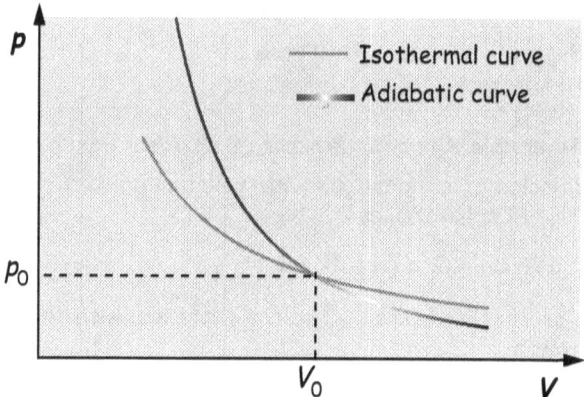

Figure 4.5 Comparison of reversible adiabatic and isothermal curves for an ideal gas.

Example

Use $C_{p,m} = 5R/2 = 20.8$ J mol^{-1}K^{-1} for a monoatomic gas assumed to behave ideally. A reversible adiabatic compression brings one mole of the gas at 300 K, 1 bar to 2 bar. From Eq. 4.29 using the ideal gas law, we get:

$$C_{V,m} \ln\left(\frac{T_2}{T_1}\right) + R \ln\left(\frac{V_2}{V_1}\right) = 0 \quad \Rightarrow \quad C_{V,m} \ln\left(\frac{T_2}{T_1}\right) + R \ln\left(\frac{T_2}{T_1}\frac{p_1}{p_2}\right) = 0$$

$$C_{p,m} \ln\left(\frac{T_2}{T_1}\right) + R \ln\left(\frac{p_1}{p_2}\right) = 0 \quad \Rightarrow \quad T_2 = T_1 \left(\frac{p_2}{p_1}\right)^{\frac{R}{C_{p,m}}} = 300 \cdot 2^{0.4} = 396 \text{ K}$$

4.4 Carnot Cycle and Entropy

On Fig. 4.6, we show two different representations of an engine Carnot cycle (of a gas). We show a p vs. V diagram as well as a T vs. S.

- 1-2   Isothermal (reversible) expansion at temperature T_h.

$$\left.\begin{array}{l} q_h = T_h (S_2 - S_1) \\ w_h = -q_h \end{array}\right\} \quad (4.31)$$

- 2-3   Adiabatic reversible expansion from T_h to T_ℓ, (isentropic)

$$\left.\begin{array}{l} dq = 0 \quad dU = C_V\, dT = dw \\ q_{23} = 0 \quad w_{23} = \displaystyle\int_{T_h}^{T_\ell} C_V\, dT \end{array}\right\} \quad (4.32)$$

4 Second Law of Thermodynamics

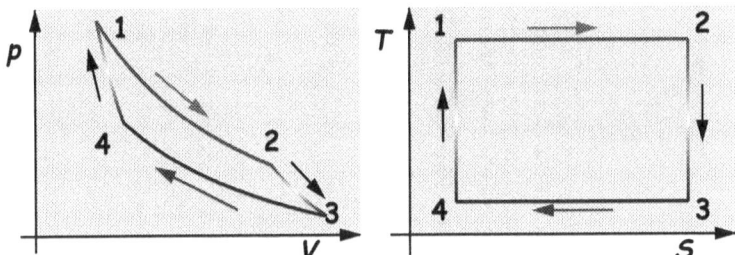

Figure 4.6 Representation of a Carnot cycle. In the direction used here, this is an engine cycle. The area inside either curve represents the work done on the system.

- 3-4 Isothermal (reversible) compression at temperature T_ℓ,

$$\left.\begin{array}{l} q_\ell = T_\ell (S_4 - S_3) = T_\ell (S_1 - S_2) \\ w_\ell = - q_\ell \end{array}\right\} \quad (4.33)$$

- 4-1 Adiabatic reversible compression from T_ℓ to T_h, (isentropic)

$$\left.\begin{array}{l} dq = 0 \quad dU = C_V \, dT = dw \\ q_{41} = 0 \quad w_{41} = \int_{T_\ell}^{T_h} C_V \, dT = - w_{23} \end{array}\right\} \quad (4.34)$$

The work done on the system is :

$$w = w_h + w_\ell = -(q_h + q_\ell) = (S_1 - S_2)(T_h - T_\ell) \quad (4.35)$$

Example

A system undergoes a Carnot cycle. The thermal reservoirs temperatures are $T_h = 1300$ K and $T_\ell = 300$ K. Let us find what the entropy change from state 1 to state 2 needed to obtain 1000 J of usable work ($w = -1000$ J). We have :

$$S_2 - S_1 = - \frac{w}{T_h - T_\ell} = \frac{1000}{1300 - 300} = 1 \text{ J K}^{-1}$$

4.5 Heat Engines, Refrigerators, Heat Pumps

4.5.1 Thermal Machines

All machines that conceptually exchange heat with at least two thermal reservoirs at different temperatures are **thermal machines**. It is of interest to evaluate the *efficiency* of such devices. The *efficiency* of a machine is the ratio of the desired energy form to the energy that had to be supplied to obtain it.

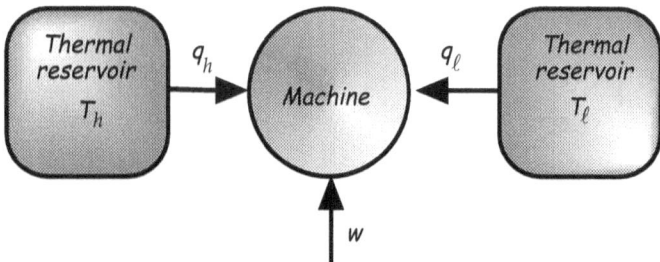

Figure 4.7 Heat and work exchanges taking place in a thermodynamic machine, in contact with two thermal reservoirs. The indicated heat and work are those received by the machine which here constitutes the system.

For refrigerators and heat pumps, this ratio is often larger than 1 and is usually called the *coefficient of performance*. In Fig. 4.7, we show the energy exchanges that take place using the sign convention mentioned earlier.

4.5.2 Efficiency of an Engine

For an engine undergoing a reversible cycle, we obtain, using the result of § 4.4:

$$\eta_e(\text{rev}) = \frac{-w}{q_h} = \frac{q_h + q_\ell}{q_h} = \frac{(S_2 - S_1)(T_h - T_\ell)}{T_h(S_2 - S_1)} = 1 - \frac{T_\ell}{T_h} \qquad (4.36)$$

In real systems, the cycle is not reversible. Using the Clausius inequality, we have:

$$\left.\begin{array}{l} \dfrac{q_h}{T_h} + \dfrac{q_\ell}{T_\ell} \leq 0 \quad \Rightarrow \quad \dfrac{q_\ell}{q_h} \leq -\dfrac{T_\ell}{T_h} \\[1em] \eta_e = 1 + \dfrac{q_\ell}{q_h} \leq 1 - \dfrac{T_\ell}{T_h} \end{array}\right\} \qquad (4.37)$$

The efficiency of a real engine is always smaller than that of a reversible engine and always smaller than 1.

Example

Let us find the maximum efficiency for a thermal engine that exchanges heat with only two thermal reservoirs at 2000 K and 300 K.

$$\eta_e(\text{rev}) = 1 - \frac{T_\ell}{T_h} = 1 - \frac{300}{2000} = 0.85$$

4.5.3 Performance of a Refrigerator

The inside of the refrigerator is the cold thermal reservoir. The outside of the refrigerator plays the role of the hot thermal reservoir. For this type of machine, q_ℓ is positive.

$$\frac{q_h}{T_h} + \frac{q_\ell}{T_\ell} \leq 0 \quad \Rightarrow \quad q_h \leq -\frac{T_h}{T_\ell} q_\ell$$

$$\text{since } q_\ell > 0 \quad \Downarrow \quad \begin{cases} q_h < 0 \\ |q_h| > q_\ell \end{cases} \quad (4.38)$$

Using the first law (Eq. 3.3), we find that:

$$w + q_h + q_\ell = 0 \quad \Rightarrow \quad \begin{cases} w = -q_h - q_\ell \\ w > 0 \end{cases} \quad (4.39)$$

The coefficient of performance, η_r, is given by:

$$\frac{q_h}{T_h} + \frac{q_\ell}{T_\ell} \leq 0 \quad \Rightarrow \quad -\frac{q_h}{q_\ell} \geq \frac{T_h}{T_\ell}$$

$$\Downarrow$$

$$\eta_r = \frac{q_\ell}{w} = \frac{q_\ell}{-q_h - q_\ell} = \frac{1}{-\frac{q_h}{q_\ell} - 1} \leq \frac{1}{\frac{T_h}{T_\ell} - 1} \quad (4.40)$$

Example

We calculate the maximum value of the coefficient of performance for a refrigerator that operates reversibly in a room at 298 K while the inside is kept at 278 K:

$$\eta_r = \frac{1}{\frac{T_h}{T_\ell} - 1} = \frac{1}{\frac{298}{278} - 1} = 13.9$$

If 100 W of work is provided, the maximum amount of heat that can be extracted from the refrigerator in 1 hour is:

$$w = 100 \cdot 3600 = 3.6 \ 10^5 \text{ J} \qquad q_\ell = \eta_r \cdot w = 13.9 \cdot 100 \cdot 3600 \simeq 5 \text{ MJ}$$

In practice this type of efficiency can never be achieved of course.
The heat of fusion of ice is 6000 J mol^{-1}. The maximum amount of water already at freezing temperature that can be transformed in ice is:

$$m = \frac{18 \ 10^{-3} \text{ kg mol}^{-1} \cdot 5 \ 10^6 \text{ J}}{6000 \text{ J mol}^{-1}} = 15 \text{ kg}$$

4.5.4 Performance of a Heat Pump

The hot source (the inside of a building) receives heat ($q_h < 0$) from the cold source ($q_\ell > 0$ for example the outside air). The coefficient of performance, η_{HP}, is :

$$\left.\begin{array}{c} \dfrac{q_h}{T_h} + \dfrac{q_\ell}{T_\ell} \leq 0 \quad \Rightarrow \quad \dfrac{q_\ell}{q_h} \geq -\dfrac{T_\ell}{T_h} \\ \Downarrow \\ \eta_{HP} = \dfrac{-q_h}{w} = \dfrac{-q_h}{-q_h - q_\ell} = \dfrac{1}{1 + \dfrac{q_\ell}{q_h}} \leq \dfrac{1}{1 - \dfrac{T_\ell}{T_h}} \end{array}\right\} \qquad (4.41)$$

The efficiency or coefficient of performance is found to be lower than that of an ideal machine operating reversibly between the same heat sources.

Example

We calculate the coefficient of performance of a heat pump that operates reversibly to warm up a house at 298 K while the outside temperature is 273 K.

$$\eta_{HP} = \dfrac{1}{1 - \dfrac{T_\ell}{T_h}} = \dfrac{1}{1 - \dfrac{273}{298}} \simeq 11.9$$

If work is provided at the rate of 100 W, the maximum amount of heat that can be released inside the house in 1 hour is :

$$w = 100 \cdot 3600 = 3.6 \; 10^5 \text{ J} \qquad q_h = \eta_{HP} \cdot w = 11.9 \cdot 100 \cdot 3600 \simeq 4.29 \text{ MJ}$$

4.6 Otto Cycle or Beau de Rochas Cycle

Figure 4.8 schematically represents this cycle. Gas engines, which operate in cars, work according to this cycle. It has four steps that are :

- 1-2   Compression of the fuel air mixture (adiabatic)
- 2-3   Combustion of the mixture at constant volume
- 3-4   Adiabatic expansion (step during which the engine provide usable work)
- 4-1   Cooling of the products of the combustion at constant volume

The exhaust of the burnt gases and the intake of the fuel air mixture are not part of the thermodynamic evaluation presented here.

4 Second Law of Thermodynamics

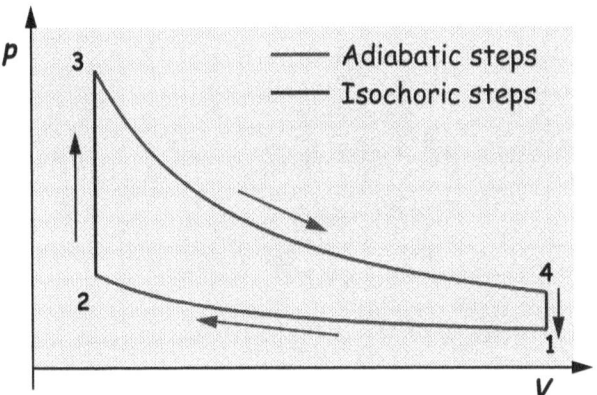

Figure 4.8 Representation of the Otto cycle (or Beau de Rochas cycle).

With a number of simplifying assumptions, we can calculate the efficiency of such a cycle. The efficiency of an engine is given by:

$$\eta_e = 1 + \frac{q_\ell}{q_h} \tag{4.42}$$

Assume:
- The fuel air mixture behaves as an ideal gas.
- The heat capacity of the gases remains constant during the cycle.
- The processes $1 \rightarrow 2$ and $3 \rightarrow 4$ are reversible adiabatic.

$$U_3 - U_2 = q_h = C_V(T_3 - T_2) \qquad U_1 - U_4 = q_\ell = C_V(T_1 - T_4) \tag{4.43}$$

$$\eta_e = 1 + \frac{q_\ell}{q_h} = 1 + \frac{T_1 - T_4}{T_3 - T_2} \tag{4.44}$$

T_1 and T_2 are on a reversible adiabatic curve, so are T_3 and T_4. By using 4.30 for a reversible adiabatic change of an ideal gas, we have:

$$\frac{T_2}{T_1} = \left(\frac{V_F}{V_I}\right)^{(\gamma-1)} \qquad \frac{T_3}{T_4} = \left(\frac{V_F}{V_I}\right)^{(\gamma-1)} \qquad \text{where } \gamma = \frac{C_{p,m}}{C_{V,m}} \tag{4.45}$$

$$\eta_e = 1 + \frac{T_1 - T_4}{T_3 - T_2} = 1 - \frac{1}{a^{(\gamma-1)}} \qquad \text{where } a = \frac{V_F}{V_I} \tag{4.46}$$

The efficiency of the cycle improves as the compression ratio a is increased.

Example

For a mixture of gases with $\gamma = 1.4$ and a compression ratio of a = 8, the efficiency is :

$$\eta_e = 1 - \frac{1}{8^{(1.4-1)}} = 0.56$$

After compression the gas mixture, initially at 300 K has reached :

$$T_2 = T_1 \left(\frac{V_F}{V_I}\right)^{(\gamma-1)} = 300 \cdot 8^{0.4} = 689 \text{ K}$$

To obtain 1000 J of useful work we must supply :

$$q_h = -\frac{w}{\eta_e} = -\frac{-1000}{0.56} \approx 1786 \text{ J}$$

Assuming we have 0.04 mol of gas mixture (1 liter of gas at 300 K) with a molar heat capacity, $C_{V,m}$ = 20.8 J mol^{-1} K^{-1}, the temperature reached by the gas at the end of the combustion is :

$$T_3 = T_2 + \frac{q_h}{n\, C_{V,m}} = 689 + \frac{1786}{0.04 \cdot 20.8} = 2836 \text{ K}$$

4.7 Stirling Cycle

This cycle has four steps and is represented in Fig. 4.9.

- 1-2 Isothermal at T_ℓ
- 2-3 At constant volume V_2
- 3-4 Isothermal at T_h
- 4-1 At constant volume V_1

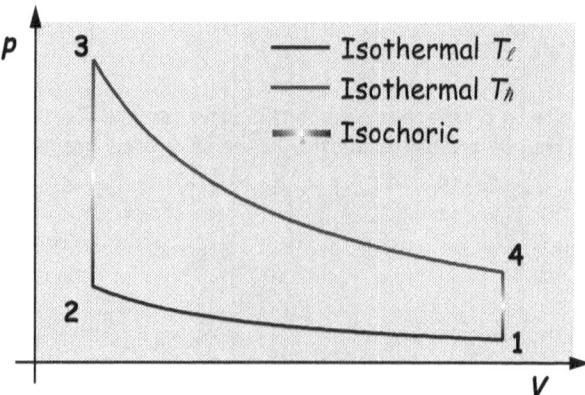

Figure 4.9 Schematic representation of the Stirling cycle.

4 Second Law of Thermodynamics

We assume that the gas is ideal.

$$\left.\begin{aligned}
&\text{1-2} \quad U_2 - U_1 = 0 \qquad\qquad q_\ell = -w_\ell = nRT_\ell \ln\left(\frac{V_2}{V_1}\right)\\
&\text{2-3} \quad U_3 - U_2 = \int_{T_\ell}^{T_h} C_V dT = q_i \quad w_{23} = 0 \\
&\text{3-4} \quad U_4 - U_3 = 0 \qquad\qquad q_h = -w_h = nRT_h \ln\left(\frac{V_1}{V_2}\right)\\
&\text{4-1} \quad U_1 - U_4 = \int_{T_h}^{T_\ell} C_V dT = -q_i \quad w_{41} = 0
\end{aligned}\right\} \qquad (4.47)$$

The work w is done on the system. We have:

$$-w = (q_\ell + q_h) = nR(T_h - T_\ell)\ln\left(\frac{V_1}{V_2}\right) \qquad (4.48)$$

The efficiency of the cycle is given by:

$$\eta_e = -\frac{w}{q_h} = 1 - \frac{T_\ell}{T_h} \qquad (4.49)$$

Example

Let us examine a Stirling cycle of 0.05 mol of N_2 behaving ideally. The thermal reservoir temperatures are 300 K and 600 K. The heat capacity of the gas is assumed to be constant and we use $C_{V,\,m} = 5\,R/2 = 20.8$ J mol^{-1}K^{-1}. In state 1, the pressure of the gas is 10^5 Pa. The volume V_2 is 1/10 of V_1. Volumes and pressures are:

$$V_1 = \frac{0.05 \cdot 8.3145 \cdot 300}{10^5} \simeq 1.247 \cdot 10^{-3} \text{ m}^3 \qquad p_2 = p_1 \frac{V_1}{V_2} = 10^6 \text{ Pa}$$

$$p_3 = p_2 \frac{T_h}{T_\ell} = 2 \cdot 10^6 \text{ Pa} \qquad\qquad p_4 = p_3 \frac{V_2}{V_1} = 2 \cdot 10^5 \text{ Pa}$$

We have the following results:

$$\left.\begin{aligned}
q_\ell = -w_\ell &= 0.05 \cdot 8.3145 \cdot 300 \ln 0.1 \simeq -287 \text{ J}\\
q_h = -w_h &= 0.05 \cdot 8.3145 \cdot 600 \ln 10 \simeq 574 \text{ J}
\end{aligned}\right\} \Rightarrow w \simeq -287 \text{ J}$$

The cyclic process produces work. The heat received by the system during step 2–3 is:

$$q_i = n\,C_{V,\,m} \int_{T_\ell}^{T_h} dT = 0.05 \cdot 20.8 \cdot (600 - 300) \simeq 312 \text{ J}$$

This amount of heat is usually obtained mostly from a heat exchanger that is heated during step 4–1 when the system receives – q_i, warming up the heat exchanger in the process. Note that steps 2–3 and 4–1 can be reversible only if the system can get in contact with an infinite number of thermal reservoirs always at the system temperature as its pressure and temperature change. If, at step 2, the system is brought in contact with the thermal reservoir at high temperature, then the heat exchange is irreversible. In this case, we can use the result of example of § 4.3.5 to find the entropy change for the global system during step 2–3 is :

$$(S_3 - S_2)_{global} = (S_3 - S_2)_{sys} + (S_3 - S_2)_{therm}$$

$$= n\, C_{V,\,m} \ln\left(\frac{T_h}{T_\ell}\right) - \frac{q_i}{T_h}$$

$$= 0.05 \cdot 20.8 \cdot \ln 2 - \frac{312}{600} = 0.20 \text{ J K}^{-1}$$

The positive result obtained agrees with the second law. If the isochoric process takes place when the system is in contact with the cold thermal reservoir then we find :

$$(S_1 - S_4)_{global} = (S_1 - S_4)_{sys} + (S_1 - S_4)_{therm}$$

$$= n\, C_{V,\,m} \ln\left(\frac{T_\ell}{T_h}\right) + \frac{q_i}{T_\ell}$$

$$= 0.05 \cdot 20.8 \cdot \ln\frac{1}{2} + \frac{312}{300} = 0.319 \text{ J K}^{-1}$$

5. Auxiliary Functions : Enthalpy, Helmholtz Energy, Gibbs Energy

5.1 Introduction

Many topics are better understood and dealt with using *auxiliary state variables* or *auxiliary functions* which turn out to be more appropriate to specific problems. These functions are simple linear combinations of the fundamental variables but have deep physical significance. They are defined using the variables :

$$\left.\begin{array}{l} U \\ S \\ V \end{array}\right\} \text{Extensive} \qquad \left.\begin{array}{l} p \\ T \end{array}\right\} \text{Intensive}$$

Auxiliary functions are known under a variety of names, most of them still in current use :

Enthalpy $\qquad\qquad H = U + p V \qquad\qquad$ (5.1)

$$\left.\begin{array}{l} \text{Helmholtz energy} \\ \text{Helmholtz function} \\ \text{Helmholtz free energy} \\ \text{Free energy} \end{array}\right\} \quad A = U - TS \qquad (5.2)$$

$$\left.\begin{array}{l} \text{Gibbs energy} \\ \text{Gibbs function} \\ \text{Gibbs free energy} \\ \text{Free enthalpy} \end{array}\right\} \quad \left\{\begin{array}{l} G = U + pV - TS \\ G = H - TS \\ G = A + pV \end{array}\right\} \quad (5.3)$$

From the way they are obtained from state functions, they are also state functions and they are extensive properties.

5.2 Closed Systems

5.2.1 Constant Volume Process (Isochoric Process)

A *process* is *isochoric* when it takes place at *constant volume*. For a change of a *closed system* at constant composition on which no *work* is done ($dV = 0$), we have :

$$dU = dq + dw = dq - p_{ext}\, dV = dq_V \Rightarrow \int_I^F dU = \int_I^F dq_V = q_V = U_F - U_I$$
(5.4)

The internal energy change corresponds to the amount of heat received by the system during a constant volume process. For a system with a *single phase*†, for a constant volume process, the variation of U can be written :

$$dU = \left(\frac{\partial U}{\partial T}\right)_V dT = C_V\, dT \Rightarrow C_V = \left(\frac{\partial U}{\partial T}\right)_V$$
(5.5)

C_V is, by definition, the **heat capacity** of the system **at constant volume**. It represents the energy needed to raise the temperature of the system by 1 K during a constant volume process.

Example

The heat capacity of a gas at constant volume is $C_V = 20.8$ J K^{-1}. During an isochoric process that causes its temperature to change from 300 K to 500 K, the heat received by the gas is equal to the change of its internal energy is :

$U_F - U_I = C_V (T_F - T_I) = 20.8 \cdot (500 - 300) = 4160$ J

5.2.2 Constant Pressure Process (Isobaric Process)

A *process* is *isobaric* when the *system pressure remains constant and equal to the external pressure* during the entire process. For a *closed system* at constant pressure, we have :

$$\left.\begin{aligned} dH &= dU + p\, dV + V\, dp = dq + dw + p\, dV + V\, dp \\ &= dq - p_{ext}\, dV + p\, dV + V\, dp \\ p &= p_{ext} \quad \text{and} \quad \text{isobaric} \Leftrightarrow dp = 0 \\ dH &= dq_p \Rightarrow \int_I^F dH = \int_I^F dq_p = q_p = H_F - H_I \end{aligned}\right\}$$
(5.6)

---

† See the definition of a phase in chapter 8.

5 Auxiliary Functions : Enthalpy, Gibbs Energy, Helmholtz Energy

The enthalpy change corresponds to the amount of heat received by the system during a constant pressure process.
For a system with a *single phase* and for an isobaric process, the variation of H is :

$$dH = \left(\frac{\partial H}{\partial T}\right)_p dT = C_p dT \quad \Rightarrow \quad C_p = \left(\frac{\partial H}{\partial T}\right)_p \tag{5.7}$$

where C_p is, by definition, the **heat capacity** of the system *at constant pressure*. It represents the energy necessary to raise the temperature of the system by 1 K during a constant pressure process.

Example
The heat capacity of a gas at constant pressure is $C_p = 60$ J K^{-1}. During an isobaric process that causes its temperature to change from 300 K to 500 K, the heat received by the gas is equal to the change of its enthalpy is :

$$H_F - H_I = C_p (T_F - T_I) = 60 \cdot (500 - 300) = 12000 \text{ J}$$

5.2.3 Monobaric Process

A **monobaric process** is a process where the *external pressure* acting on the system *remains constant*. The system pressure is in general ill defined during the process, but, in the initial and final states, the system pressure is assumed to be equal to the external pressure, $p_{ext} = p_I = p$. Applying the first law to such a system and assuming the only work done on the system is due to volume change, we have :

$$\left.\begin{array}{c} U_F - U_I = w + q_p = - p_{ext}(V_F - V_I) + q_p \\ \Downarrow \\ (U_F + p_{ext} V_F) - (U_I + p_{ext} V_I) = q_p = H_F - H_I \end{array}\right\} \tag{5.8}$$

We find that the enthalpy change corresponds to the amount of heat received by the system during a *monobaric process* (constant external pressure).

Example
A cylinder closed by a piston contains 0.1 mole of an ideal gas ($C_{p,m} = 29.1$ J mol^{-1} K^{-1}) at 500 K. This system is placed in contact with a thermal reservoir at 300 K while the external pressure remains constant. An irreversible heat exchange takes place. The amount of heat received by the system is :

$$H_F - H_I = nC_{p,m}(T_F - T_I) = 0.1 \cdot 29.1 \cdot (300 - 500) = -582 \text{ J}$$

The system transfers heat to the thermal reservoir. Note that work is also done on the system.

5.3 Characteristic Variables. Fundamental Equations. Open Systems. Systems with Chemical Reactions

5.3.1 Generalities

A function of state can be expressed in terms of two variables together with the amount (number of moles) of each species of the system (a subscript i indicates the amount refers to species i). For an open system or a system where chemical reactions can take place, we have:

$$U = f(X, Y, ..., n_i, ...) \tag{5.9}$$

where X and Y are state variables which can be selected amongst p, V, T, H, S, A, G, C_p, C_V. From the expression for the differential of the internal energy of a closed system (Eq. 4.13), we have:

$$dU = -p\,dV + T\,dS \tag{4.13}$$

The variables V and S are called characteristic variables of the internal energy. We write $U(V, S, ..., n_i, ...)$.

5.3.2 Internal Energy

For systems where chemical reactions can take place or which are open systems (matter can enter or leave the system), the differential of the internal energy is:

$$dU = \left(\frac{\partial U}{\partial V}\right)_{S, n_i} dV + \left(\frac{\partial U}{\partial S}\right)_{V, n_i} dS + \sum_i \left(\frac{\partial U}{\partial n_i}\right)_{V, S, n_j} dn_i \tag{5.10}$$

since dU is an exact differential. The subscript n_i indicates that all mole numbers are kept constant, the subscript n_j indicates that all mole numbers except n_i stay constant. This expression must remain valid for a *closed system* in which *no chemical reaction* takes place. All of dn_i then are zero and expression 5.10 must be identical to 4.13. Hence:

$$\left.\begin{array}{l}\left(\dfrac{\partial U}{\partial V}\right)_{S, n_i} = -p \qquad \left(\dfrac{\partial U}{\partial S}\right)_{V, n_i} = T \\[1em] dU = -p\,dV + T\,dS + \sum_i \left(\dfrac{\partial U}{\partial n_i}\right)_{V, S, n_j} dn_i\end{array}\right\} \tag{5.11}$$

Example

The internal energy of an ideal gas depends only on its temperature. Assume $C_{V, m}$ is constant. For a reversible adiabatic process (*constant entropy*) of a closed system, we have for n moles:

$$dU = -p\,dV \quad\Rightarrow\quad nC_{V, m}\,dT = -\frac{nRT}{V}dV \quad\Rightarrow\quad \frac{dT}{T} = -\frac{R}{C_{V, m}}\frac{dV}{V}$$

5 Auxiliary Functions : Enthalpy, Gibbs Energy, Helmholtz Energy

By integration of this differential equation, we obtain :

$$\ln \frac{T_2}{T_1} = -\frac{R}{C_{V,m}} \ln \frac{V_2}{V_1} \Rightarrow T_2 = T_1 \left(\frac{V_2}{V_1}\right)^{-\frac{R}{C_{V,m}}}$$

Using $C_{V,m} = 5\,R/2$, we find how the gas cools when its volume doubles and $T_1 = 300$ K :

$$T_2 = 300 \cdot 2^{-0.4} \simeq 227 \text{ K}$$

5.3.3 Enthalpy

Using Eq. 5.1 and 5.11, we can write :

$$\left. \begin{array}{l} H = U + pV \Rightarrow dH = dU + p\,dV + V\,dp \\ dH = T\,dS + V\,dp + \sum_i \left(\frac{\partial U}{\partial n_i}\right)_{V,S,n_j} dn_i \end{array} \right\} \quad (5.12)$$

The variables S and p are called characteristic variables of the enthalpy. We write $H(S, p, ..., n_i, ...)$. Since enthalpy is a function of state, we also have :

$$dH = \left(\frac{\partial H}{\partial S}\right)_{p, n_i} dS + \left(\frac{\partial H}{\partial p}\right)_{S, n_i} dp + \sum_i \left(\frac{\partial H}{\partial n_i}\right)_{p, S, n_j} dn_i \quad (5.13)$$

We can see that we have the following relations :

$$\left. \begin{array}{l} \left(\frac{\partial H}{\partial S}\right)_{p, n_i} = T \quad \left(\frac{\partial H}{\partial p}\right)_{S, n_i} = V \\ \left(\frac{\partial H}{\partial n_i}\right)_{p, S, n_j} = \left(\frac{\partial U}{\partial n_i}\right)_{V, S, n_j} = \mu_i \end{array} \right\} \quad (5.14)$$

The fact that the last two partial derivatives are equal, leads us to represent them by a unique symbol, μ_i.

> The quantity μ_i is named the **chemical potential** of species i. The chemical potential is an **intensive** variable.

Example

The enthalpy of an ideal gas depends only on temperature. Assume $C_{p,m}$ is constant. For a closed system undergoing a change at *constant entropy*, we have for n moles :

$$dH = V\,dp \Rightarrow n\,C_{p,m}\,dT = \frac{nRT}{p}\,dp \Rightarrow \frac{dT}{T} = \frac{R}{C_{p,m}}\,\frac{dp}{p}$$

By integration of this differential equation, we obtain :

$$\ln \frac{T_2}{T_1} = \frac{R}{C_{p,m}} \ln \frac{p_2}{p_1} \Rightarrow T_2 = T_1 \left(\frac{p_2}{p_1}\right)^{\frac{R}{C_{p,m}}}$$

Using $C_{p,m} = 7\,R/2$, we find how the gas heats up when its pressure doubles and $T_1 = 300$ K :
$$T_2 = 300 \cdot 2^{0.286} = 366 \text{ K}$$

5.3.4 Helmholtz Energy (Helmholtz Function, Free Energy)

Using Eq. 5.2 and 5.11, we obtain :
$$A = U - TS \Rightarrow dA = dU - TdS - SdT$$
$$dA = -p\,dV - S\,dT + \sum_i \left(\frac{\partial U}{\partial n_i}\right)_{V, S, n_j} dn_i \qquad (5.15)$$

The variables V and T are called characteristic variables of the Helmholtz energy. We write $A(V, T, ..., n_i, ...)$. Since the Helmholtz energy (free energy) is a state function, we have :

$$dA = \left(\frac{\partial A}{\partial V}\right)_{T, n_i} dV + \left(\frac{\partial A}{\partial T}\right)_{V, n_i} dT + \sum_i \left(\frac{\partial A}{\partial n_i}\right)_{V, T, n_j} dn_i \qquad (5.16)$$

We have the following relations :
$$\left(\frac{\partial A}{\partial V}\right)_{T, n_i} = -p \qquad \left(\frac{\partial A}{\partial T}\right)_{V, n_i} = -S$$
$$\left(\frac{\partial A}{\partial n_i}\right)_{V, T, n_j} = \left(\frac{\partial U}{\partial n_i}\right)_{V, S, n_j} = \mu_i \qquad (5.17)$$

Example
One mole of an ideal gas undergoes a reversible volume change at constant temperature. At 300 K, the gas volume of one mole of gas decreases by a factor of two during the process. We find :
$$A_2 - A_1 = \int_{V_1}^{V_2} -p\,dV = -\int_{V_1}^{V_2} \frac{nRT}{V}dV = -n\,R\,T\,\ln\frac{V_2}{V_1} = -8.3145 \cdot 300 \cdot \ln\frac{1}{2} \approx 1729 \text{ J}$$

The change in the Helmholtz energy corresponds to the (reversible) work done on the system.

5.3.5 Gibbs Energy (Gibbs Function, Free Enthalpy)

$$G = H - TS \Rightarrow dG = dH - TdS - SdT$$
$$dG = V\,dp - S\,dT + \sum_i \left(\frac{\partial U}{\partial n_i}\right)_{V, S, n_j} dn_i \qquad (5.18)$$

The variables p and T are called characteristic variables of the Gibbs energy. We write $G(p, T, ..., n_i, ...)$.

5 Auxiliary Functions : Enthalpy, Gibbs Energy, Helmholtz Energy

Since the Gibbs energy (free enthalpy) is a state function, we have :

$$dG = \left(\frac{\partial G}{\partial p}\right)_{T, n_i} dp + \left(\frac{\partial G}{\partial T}\right)_{p, n_i} dT + \sum_i \left(\frac{\partial G}{\partial n_i}\right)_{p, T, n_j} dn_i \quad (5.19)$$

Comparing 5.18 and 5.19 leads to :

$$\left. \begin{array}{l} \left(\dfrac{\partial G}{\partial T}\right)_{p, n_i} = -S \qquad \left(\dfrac{\partial G}{\partial p}\right)_{T, n_i} = V \\[6pt] \left(\dfrac{\partial G}{\partial n_i}\right)_{p, T, n_j} = \left(\dfrac{\partial U}{\partial n_i}\right)_{V, S, n_j} = \mu_i \end{array} \right\} \quad (5.20)$$

Example

One mole of an ideal gas undergoes a reversible pressure change at constant temperature. At 300 K, the pressure of one mole of gas increases by a factor of two during the process. The change in the Gibbs energy is :

$$G_2 - G_1 = \int_{p_1}^{p_2} V \, dp = \int_{p_1}^{p_2} \frac{nRT}{p} dp = nRT \ln \frac{p_2}{p_1} = 8.3145 \cdot 300 \cdot \ln 2 \simeq 1729 \text{ J}$$

The change in the Gibbs energy corresponds to the (reversible) work done on the system.

5.3.6 Chemical Potential. Summary

$$\mu_i = \left(\frac{\partial U}{\partial n_i}\right)_{V, S, n_j} = \left(\frac{\partial H}{\partial n_i}\right)_{p, S, n_j} = \left(\frac{\partial A}{\partial n_i}\right)_{T, V, n_j} = \left(\frac{\partial G}{\partial n_i}\right)_{p, T, n_j} \quad (5.21)$$

As we will see in chapter 6, the derivative of an extensive variable with respect to another extensive variable is an intensive variable. The *chemical potential* is an *intensive variable*. In the differential expressions of any extensive variable, an extensive variable (V, S, n_i) is always associated with an intensive variable (p, T, μ_i).

$$\left. \begin{array}{l} dU = -p\,dV + T\,dS + \sum_i \mu_i \, dn_i \\[6pt] dH = T\,dS + V\,dp + \sum_i \mu_i \, dn_i \\[6pt] dA = -p\,dV - S\,dT + \sum_i \mu_i \, dn_i \\[6pt] dG = V\,dp - S\,dT + \sum_i \mu_i \, dn_i \end{array} \right\} \quad (5.22)$$

$$T = \left(\frac{\partial U}{\partial S}\right)_{V, n_i} = \left(\frac{\partial H}{\partial S}\right)_{p, n_i}$$

$$p = -\left(\frac{\partial U}{\partial V}\right)_{S, n_i} = -\left(\frac{\partial A}{\partial V}\right)_{T, n_i}$$

$$V = \left(\frac{\partial H}{\partial p}\right)_{S, n_i} = \left(\frac{\partial G}{\partial p}\right)_{T, n_i}$$

$$S = -\left(\frac{\partial A}{\partial T}\right)_{V, n_i} = -\left(\frac{\partial G}{\partial T}\right)_{p, n_i}$$

(5.23)

Additionally, we can express the Helmholtz energy and the Gibbs energy in the following form:

$$A = U - TS = U + T\left(\frac{\partial A}{\partial T}\right)_{V, n_i}$$

$$G = H - TS = H + T\left(\frac{\partial G}{\partial T}\right)_{p, n_i}$$

(5.24)

This formalism can then be used to express the variation of A and G with temperature.

$$\left[\frac{\partial}{\partial T}\left(\frac{A}{T}\right)\right]_{V, n_i} = \frac{1}{T}\left(\frac{\partial A}{\partial T}\right)_{V, n_i} - \frac{A}{T^2} = -\frac{U}{T^2}$$

$$\left[\frac{\partial}{\partial T}\left(\frac{G}{T}\right)\right]_{p, n_i} = \frac{1}{T}\left(\frac{\partial G}{\partial T}\right)_{p, n_i} - \frac{G}{T^2} = -\frac{H}{T^2}$$

(5.25)

Equations 5.25 are the **Gibbs-Helmholtz equations**.

Example

One mole of an ideal gas undergoes a temperature change at constant volume. We have:

$$\int_{T_1}^{T_2} d\left(\frac{A}{T}\right) = -n\,C_{V, m}\int_{T_1}^{T_2} \frac{T - T_0}{T^2}\,dT$$

$$\Downarrow$$

$$\frac{A_2}{T_2} - \frac{A_1}{T_1} = -n\,C_{V, m}\left[\ln\frac{T_2}{T_1} + T_0\left(\frac{1}{T_2} - \frac{1}{T_1}\right)\right]$$

For a change at constant pressure, we have:

$$\int_{T_1}^{T_2} d\left(\frac{G}{T}\right) = -n\,C_{p, m}\int_{T_1}^{T_2} \frac{T - T_0}{T^2}\,dT$$

$$\Downarrow$$

$$\frac{G_2}{T_2} - \frac{G_1}{T_1} = -n\,C_{p, m}\left[\ln\frac{T_2}{T_1} + T_0\left(\frac{1}{T_2} - \frac{1}{T_1}\right)\right]$$

5 Auxiliary Functions: Enthalpy, Gibbs Energy, Helmholtz Energy

5.4 Maxwell's Relations

By applying Schwarz theorem to the differential expressions obtained in § 5.3 (See § 1.6.3 – the partial derivative of a function with respect to two variables is independent of the order in which the derivation is carried out), equations 5.22 and 5.23 give:

$$\left.\begin{array}{r}\left[\dfrac{\partial}{\partial V}\left(\dfrac{\partial U}{\partial S}\right)_{V,n_i}\right]_{S,n_i} = \left(\dfrac{\partial T}{\partial V}\right)_{S,n_i} \\ \left[\dfrac{\partial}{\partial S}\left(\dfrac{\partial U}{\partial V}\right)_{S,n_i}\right]_{V,n_i} = -\left(\dfrac{\partial p}{\partial S}\right)_{V,n_i}\end{array}\right\} \Rightarrow \left(\dfrac{\partial T}{\partial V}\right)_{S,n_i} = -\left(\dfrac{\partial p}{\partial S}\right)_{V,n_i}$$

(5.26)

We can operate in the same way with other differential forms: dH, dA, dG.

$$\left.\begin{array}{r}\left[\dfrac{\partial}{\partial p}\left(\dfrac{\partial H}{\partial S}\right)_{p,n_i}\right]_{S,n_i} = \left(\dfrac{\partial T}{\partial p}\right)_{S,n_i} \\ \left[\dfrac{\partial}{\partial S}\left(\dfrac{\partial H}{\partial p}\right)_{S,n_i}\right]_{p,n_i} = \left(\dfrac{\partial V}{\partial S}\right)_{p,n_i}\end{array}\right\} \Rightarrow \left(\dfrac{\partial T}{\partial p}\right)_{S,n_i} = \left(\dfrac{\partial V}{\partial S}\right)_{p,n_i}$$

(5.27)

$$\left.\begin{array}{r}\left[\dfrac{\partial}{\partial V}\left(\dfrac{\partial A}{\partial T}\right)_{V,n_i}\right]_{T,n_i} = -\left(\dfrac{\partial S}{\partial V}\right)_{T,n_i} \\ \left[\dfrac{\partial}{\partial T}\left(\dfrac{\partial A}{\partial V}\right)_{T,n_i}\right]_{V,n_i} = -\left(\dfrac{\partial p}{\partial T}\right)_{V,n_i}\end{array}\right\} \Rightarrow \left(\dfrac{\partial S}{\partial V}\right)_{T,n_i} = \left(\dfrac{\partial p}{\partial T}\right)_{V,n_i}$$

(5.28)

$$\left.\begin{array}{r}\left[\dfrac{\partial}{\partial p}\left(\dfrac{\partial G}{\partial T}\right)_{p,n_i}\right]_{T,n_i} = -\left(\dfrac{\partial S}{\partial p}\right)_{T,n_i} \\ \left[\dfrac{\partial}{\partial T}\left(\dfrac{\partial G}{\partial p}\right)_{T,n_i}\right]_{p,n_i} = \left(\dfrac{\partial V}{\partial T}\right)_{p,n_i}\end{array}\right\} \Rightarrow -\left(\dfrac{\partial S}{\partial p}\right)_{T,n_i} = \left(\dfrac{\partial V}{\partial T}\right)_{p,n_i}$$

(5.29)

Using n_i as one of the variables, we have:

$$\left.\begin{array}{r}\left[\dfrac{\partial}{\partial n_i}\left(\dfrac{\partial G}{\partial T}\right)_{p,n_i,n_j}\right]_{T,p,n_j}=-\left(\dfrac{\partial S}{\partial n_i}\right)_{T,p,n_j}\\[2ex]\left[\dfrac{\partial}{\partial T}\left(\dfrac{\partial G}{\partial n_i}\right)_{T,p,n_j}\right]_{p,n_i,n_j}=\left(\dfrac{\partial \mu_i}{\partial T}\right)_{p,n_i,n_j}\end{array}\right\}\Rightarrow\left(\dfrac{\partial \mu_i}{\partial T}\right)_{p,n_i,n_j}=-\left(\dfrac{\partial S}{\partial n_i}\right)_{T,p,n_j}$$

(5.30)

$$\left.\begin{array}{r}\left[\dfrac{\partial}{\partial n_i}\left(\dfrac{\partial G}{\partial p}\right)_{T,n_i,n_j}\right]_{T,p,n_j}=\left(\dfrac{\partial V}{\partial n_i}\right)_{T,p,n_j}\\[2ex]\left[\dfrac{\partial}{\partial p}\left(\dfrac{\partial G}{\partial n_i}\right)_{T,p,n_j}\right]_{T,n_i,n_j}=\left(\dfrac{\partial \mu_i}{\partial p}\right)_{T,n_i,n_j}\end{array}\right\}\Rightarrow\left(\dfrac{\partial \mu_i}{\partial p}\right)_{T,n_i,n_j}=\left(\dfrac{\partial V}{\partial n_i}\right)_{T,p,n_j}$$

(5.31)

$$\left.\begin{array}{r}\left(\dfrac{\partial \mu_i}{\partial T}\right)_{V,n_i,n_j}=-\left(\dfrac{\partial S}{\partial n_i}\right)_{T,V,n_j}\\[2ex]\left(\dfrac{\partial \mu_i}{\partial p}\right)_{S,n_i,n_j}=\left(\dfrac{\partial V}{\partial n_i}\right)_{S,p,n_j}\end{array}\right\}$$

(5.32)

Using n_i as one of the variables and the fact that U is an exact differential, one finds :

$$\left.\begin{array}{r}\left(\dfrac{\partial \mu_i}{\partial S}\right)_{V,n_i,n_j}=\left(\dfrac{\partial T}{\partial n_i}\right)_{V,S,n_j}\\[2ex]\left(\dfrac{\partial \mu_i}{\partial V}\right)_{S,n_i,n_j}=-\left(\dfrac{\partial p}{\partial n_i}\right)_{V,S,n_j}\\[2ex]\left(\dfrac{\partial \mu_i}{\partial n_j}\right)_{V,S,n_{k\ne j}}=\left(\dfrac{\partial \mu_j}{\partial n_i}\right)_{V,S,n_{k\ne i}}\end{array}\right\}$$

(5.33)

Example

Using a Maxwell relation and the van der Waals equation of state :

$$\left(\dfrac{\partial S}{\partial V}\right)_{T,n}=\left(\dfrac{\partial p}{\partial T}\right)_{V,n}\qquad p=\dfrac{nRT}{V-nb}-a\dfrac{n^2}{V^2}$$

We get for the entropy change in an isothermal process :

$$\left(\dfrac{\partial p}{\partial T}\right)_{V,n}=\dfrac{nR}{V-nb}\Rightarrow S_2-S_1=\int_{V_1}^{V_2}\dfrac{nR}{V-nb}dV=nR\ln\left(\dfrac{V_2-nb}{V_1-nb}\right)$$

5 Auxiliary Functions : Enthalpy, Gibbs Energy, Helmholtz Energy

5.5 Thermodynamic Equation of State

5.5.1 General Case

Consider a *closed system* where no chemical reaction takes place and comprising only one phase.

$$dU = -p\,dV + T\,dS \tag{4.13}$$

where U is expressed as a function of V and S. We can also express the differential of U using V and T as variables. For that, we express S as a function of V and T:

$$dS = \left(\frac{\partial S}{\partial V}\right)_T dV + \left(\frac{\partial S}{\partial T}\right)_V dT \tag{5.34}$$

$$dU = \left\{-p + T\left(\frac{\partial S}{\partial V}\right)_T\right\} dV + T\left(\frac{\partial S}{\partial T}\right)_V dT \tag{5.35}$$

This expression leads to two relations. Comparing with Eq. 5.5, we obtain:

$$C_V = T\left(\frac{\partial S}{\partial T}\right)_V \tag{5.36}$$

We also get (5.28):

$$\left(\frac{\partial U}{\partial V}\right)_T = T\left(\frac{\partial S}{\partial V}\right)_T - p = T\left(\frac{\partial p}{\partial T}\right)_V - p \tag{5.37}$$

This formula is known as the **thermodynamic equation of state**. Apply this relation to the equation of state of an ideal gas.

$$p = \frac{nRT}{V} \Rightarrow \left(\frac{\partial p}{\partial T}\right)_V = \frac{nR}{V} \Rightarrow \left(\frac{\partial U}{\partial V}\right)_T = T\frac{nR}{V} - p = 0 \tag{5.38}$$

The internal energy of an ideal gas is independent of its volume.

Example

Using Eq. 5.37, we can find how the internal energy of a van der Waals gas (See chapter 7) changes with volume :

$$p = \frac{nRT}{V-nb} - a\frac{n^2}{V^2} \Rightarrow \left(\frac{\partial p}{\partial T}\right)_{V,n} = \frac{nR}{V-nb} \Rightarrow T\left(\frac{\partial p}{\partial T}\right)_{V,n} - p = a\frac{n^2}{V^2}$$

The differential of the internal energy of a van der Waals gas is therefore :

$$dU = n\,C_{V,m}\,dT + a\frac{n^2}{V^2}dV$$

At constant temperature, the volume of one mole of nitrogen changes from 2 liters to 1 liter. Using $a = 0.137$ Pa m^6 mol^{-2}, we find that the internal energy change is :

$$U_2 - U_1 = a n^2 \int_{V_1}^{V_2} \frac{dV}{V^2} = a n^2\left(\frac{1}{V_1} - \frac{1}{V_2}\right) = 0.137 \cdot \left(\frac{1}{10^{-3}} - \frac{1}{2\cdot 10^{-3}}\right) = 68.5 \text{ J}$$

Similarly expressing H as a function of p and T, we have:
$$dH = TdS + Vdp \tag{5.39}$$
The differential of entropy, as a function of the variables that we are interested in, is:
$$dS = \left(\frac{\partial S}{\partial p}\right)_T dp + \left(\frac{\partial S}{\partial T}\right)_p dT \tag{5.40}$$
Use dS in the expression for dH. We obtain:
$$dH = \left\{V + T\left(\frac{\partial S}{\partial p}\right)_T\right\} dp + T\left(\frac{\partial S}{\partial T}\right)_p dT \tag{5.41}$$
We find using 5.29:
$$C_p = T\left(\frac{\partial S}{\partial T}\right)_p \tag{5.42}$$
$$\left(\frac{\partial H}{\partial p}\right)_T = V + T\left(\frac{\partial S}{\partial p}\right)_T = V - T\left(\frac{\partial V}{\partial T}\right)_p \tag{5.43}$$
Let us get the partial derivative of the enthalpy of an ideal gas with respect to pressure at constant temperature.
$$V = \frac{nRT}{p} \Rightarrow \left(\frac{\partial V}{\partial T}\right)_{p,n} = \frac{nR}{p} \Rightarrow \left(\frac{\partial H}{\partial p}\right)_{T,n} = V - T\frac{nR}{p} = 0 \tag{5.44}$$
The enthalpy of an ideal gas is independent of its pressure.

Example

Using the isobaric coefficient of thermal expansion we have:
$$\alpha = \frac{1}{V}\left(\frac{\partial V}{\partial T}\right)_p \Rightarrow \left(\frac{\partial H}{\partial p}\right)_T = V(1 - \alpha T)$$
For one mole of chloroform at 298 K, $\alpha = 1.33 \cdot 10^{-3}$ K^{-1}, $V_m = 80.5 \cdot 10^{-6}$ m^3, we can find the enthalpy change for 10^5 Pa pressure change.
$$\Delta H = 80.5 \cdot 10^{-6} \cdot (1 - 1.33 \cdot 10^{-3} \cdot 298) \cdot 10^5 = 4.86 \text{ J}$$

5.5.2 Equation of State for an Ideal Gas

An **ideal gas** has its internal energy independent of its volume [$U(V,T) = U(T)$] and its enthalpy independent of its pressure [$H(p,T) = H(T)$]. Eq. 5.1 and 5.37 imply:

$$\left.\begin{array}{l} pV = H(T) - U(T) = f(T) \\ \left(\frac{\partial U}{\partial V}\right)_T = 0 = T\left(\frac{\partial p}{\partial T}\right)_V - p \end{array}\right\} \Rightarrow \left\{\begin{array}{l} T\frac{1}{V}\frac{df}{dT} = p \\ \frac{df}{dT} = \frac{pV}{T} = \frac{f}{T} \\ f = \text{Constant} \cdot T \end{array}\right. \tag{5.45}$$

5 Auxiliary Functions : Enthalpy, Gibbs Energy, Helmholtz Energy

$$\frac{pV}{T} = \text{Constant} \Rightarrow \begin{cases} \dfrac{pV_m}{T} = R & \text{for one mole} \\ \dfrac{pV}{T} = nR & \text{for } n \text{ moles} \end{cases} \quad (5.46)$$

The constant R (8.314472 J mol^{-1} K^{-1}) is the **gas constant**. The volume of one mole of gas is its **molar volume**, V_m.

5.6 Properties of C_p and C_V

5.6.1 Relation between C_p and C_V

As a starting point, use Eq. 5.36 and 5.42 :

$$\left. \begin{array}{l} C_V = T \left(\dfrac{\partial S}{\partial T}\right)_V \\[6pt] C_p = T \left(\dfrac{\partial S}{\partial T}\right)_p \end{array} \right\} \Rightarrow C_p - C_V = T \left\{ \left(\dfrac{\partial S}{\partial T}\right)_p - \left(\dfrac{\partial S}{\partial T}\right)_V \right\} \quad (5.47)$$

From :

$$dS = \left(\dfrac{\partial S}{\partial V}\right)_T dV + \left(\dfrac{\partial S}{\partial T}\right)_V dT \quad (5.34)$$

$$dS = \left(\dfrac{\partial S}{\partial p}\right)_T dp + \left(\dfrac{\partial S}{\partial T}\right)_p dT \quad (5.40)$$

The differential of the volume, V, considered as a function of p and T is :

$$dV = \left(\dfrac{\partial V}{\partial p}\right)_T dp + \left(\dfrac{\partial V}{\partial T}\right)_p dT \quad (5.48)$$

Using 5.48 in 5.34, the differential of S expressed as a function of dp and dT is :

$$\begin{aligned} dS &= \left(\dfrac{\partial S}{\partial V}\right)_T \left\{ \left(\dfrac{\partial V}{\partial p}\right)_T dp + \left(\dfrac{\partial V}{\partial T}\right)_p dT \right\} + \left(\dfrac{\partial S}{\partial T}\right)_V dT \\ &= \left(\dfrac{\partial S}{\partial V}\right)_T \left(\dfrac{\partial V}{\partial p}\right)_T dp + \left\{ \left(\dfrac{\partial V}{\partial T}\right)_p \left(\dfrac{\partial S}{\partial V}\right)_T + \left(\dfrac{\partial S}{\partial T}\right)_V \right\} dT \end{aligned} \quad (5.49)$$

Identifying the coefficients of dT in 5.40 and 5.49, we have :

$$\left. \begin{array}{l} \left(\dfrac{\partial S}{\partial T}\right)_p = \left(\dfrac{\partial V}{\partial T}\right)_p \left(\dfrac{\partial S}{\partial V}\right)_T + \left(\dfrac{\partial S}{\partial T}\right)_V \\[4pt] \Downarrow \\[4pt] \left(\dfrac{\partial S}{\partial T}\right)_p - \left(\dfrac{\partial S}{\partial T}\right)_V = \left(\dfrac{\partial V}{\partial T}\right)_p \left(\dfrac{\partial S}{\partial V}\right)_T \end{array} \right\} \quad (5.50)$$

Use 5.28:

$$C_p - C_V = T\left(\frac{\partial V}{\partial T}\right)_p \left(\frac{\partial S}{\partial V}\right)_T = T\left(\frac{\partial V}{\partial T}\right)_p \left(\frac{\partial p}{\partial T}\right)_V \quad (5.51)$$

For an *ideal gas*, we find:

$$\left.\begin{array}{l}\left(\dfrac{\partial V}{\partial T}\right)_p = \dfrac{nR}{p} \\ \left(\dfrac{\partial p}{\partial T}\right)_V = \dfrac{nR}{V}\end{array}\right\} \Rightarrow \quad C_p - C_V = T\dfrac{nR}{V}\dfrac{nR}{p} = nR \quad (5.52)$$

We can also express 5.51 using:

$$\alpha = \frac{1}{V}\left(\frac{\partial V}{\partial T}\right)_{p,n} \qquad \kappa = -\frac{1}{V}\left(\frac{\partial V}{\partial p}\right)_{T,n} \quad (1.5, 1.6)$$

Using properties of partial derivatives, we have:

$$\left(\frac{\partial S}{\partial p}\right)_T = \left(\frac{\partial S}{\partial V}\right)_T \left(\frac{\partial V}{\partial p}\right)_T \Leftrightarrow \left(\frac{\partial S}{\partial V}\right)_T = \frac{\left(\dfrac{\partial S}{\partial p}\right)_T}{\left(\dfrac{\partial V}{\partial p}\right)_T} \quad (5.53)$$

$$C_p - C_V = T\left(\frac{\partial V}{\partial T}\right)_p \left(\frac{\partial S}{\partial V}\right)_T = T\left(\frac{\partial V}{\partial T}\right)_p \frac{\left(\dfrac{\partial S}{\partial p}\right)_T}{\left(\dfrac{\partial V}{\partial p}\right)_T} \quad (5.54)$$

Using 1.5 and 1.6, we find:

$$\left.\begin{array}{l}\dfrac{\left(\dfrac{\partial V}{\partial T}\right)_p}{\left(\dfrac{\partial V}{\partial p}\right)_T} = \left(\dfrac{\alpha V}{-\kappa V}\right) = -\dfrac{\alpha}{\kappa} \\ \left(\dfrac{\partial S}{\partial p}\right)_T = -\left(\dfrac{\partial V}{\partial T}\right)_p = -\alpha V\end{array}\right\} \Rightarrow C_p - C_V = \dfrac{\alpha^2 VT}{\kappa} \quad (5.55)$$

5.6.2 Variation of C_V with Volume and of C_p with Pressure

Using Schwarz theorem, we have:

$$\left.\begin{array}{l}\left(\dfrac{\partial C_V}{\partial V}\right)_T = \left[\dfrac{\partial}{\partial V}\left(\dfrac{\partial U}{\partial T}\right)_V\right]_T = \left[\dfrac{\partial}{\partial T}\left(\dfrac{\partial U}{\partial V}\right)_T\right]_V \\ = \left[\dfrac{\partial}{\partial T}\left\{T\left(\dfrac{\partial p}{\partial T}\right)_V - p\right\}\right]_V\end{array}\right\} \quad (5.56)$$

5 Auxiliary Functions : Enthalpy, Gibbs Energy, Helmholtz Energy

$$\left(\frac{\partial C_V}{\partial V}\right)_T = \left(\frac{\partial p}{\partial T}\right)_V + T\left(\frac{\partial^2 p}{\partial T^2}\right)_V - \left(\frac{\partial p}{\partial T}\right)_V \\ = T\left(\frac{\partial^2 p}{\partial T^2}\right)_V \Biggr\} \quad (5.57)$$

Calculating the second derivative, we find that the heat capacity at constant volume of an ideal gas is independent of the volume of the system.

Example
For a gas obeying the van der Waals equation we have :

$$p = \frac{nRT}{V-nb} - a\frac{n^2}{V^2} \Rightarrow \left(\frac{\partial p}{\partial T}\right)_V = \frac{nR}{V-nb} \Rightarrow \left(\frac{\partial^2 p}{\partial T^2}\right)_V = 0$$

We find that the heat capacity at constant volume is independent of the gas volume.

$$\left(\frac{\partial C_p}{\partial p}\right)_T = \left[\frac{\partial}{\partial p}\left(\frac{\partial H}{\partial T}\right)_p\right]_T = \left[\frac{\partial}{\partial T}\left(\frac{\partial H}{\partial p}\right)_T\right]_p \\ = \left[\frac{\partial}{\partial T}\left\{V - T\left(\frac{\partial V}{\partial T}\right)_p\right\}\right]_p \Biggr\} \quad (5.58)$$

$$\left(\frac{\partial C_p}{\partial p}\right)_T = \left(\frac{\partial V}{\partial T}\right)_p - T\left(\frac{\partial^2 V}{\partial T^2}\right)_p - \left(\frac{\partial V}{\partial T}\right)_p \\ = -T\left(\frac{\partial^2 V}{\partial T^2}\right)_p \Biggr\} \quad (5.59)$$

Similarly, the heat capacity of an ideal gas at constant pressure is independent of the system pressure.

Example
Let us find the change of C_p with pressure for a gas obeys the following equation of state :

$$p\{V - nb(T)\} = nRT \Rightarrow \left(\frac{\partial V}{\partial T}\right)_{p,n} = \frac{nR}{p} + n\frac{db(T)}{dT} \Rightarrow \left(\frac{\partial^2 V}{\partial T^2}\right)_p = n\frac{d^2 b(T)}{dT^2}$$

$$\left(\frac{\partial C_p}{\partial p}\right)_T = -nT\frac{d^2 b(T)}{dT^2}$$

5.7 Physical Meaning of the Auxiliary Functions

5.7.1 Helmholtz Energy (Helmholtz Function, Free Energy)

Consider a finite change from state I to state F of a *closed system* which can be in contact with a thermal reservoir at temperature T_{therm} during part of the process. System and thermal reservoir together make up an adiabatic closed system, the global system (Fig. 5.1).

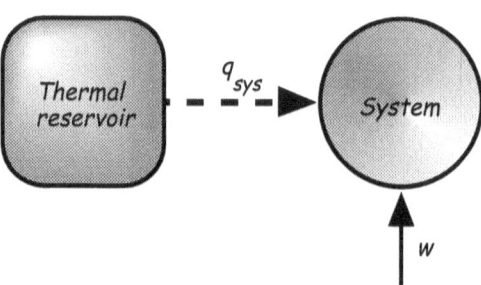

Figure 5.1 Closed system in contact with a thermal reservoir during part of the process.

$$q_{therm} = -q_{sys} = -(U_F - U_I)_{sys} + w$$
$$\Downarrow$$
$$(S_F - S_I)_{therm} = \frac{q_{therm}}{T_{therm}} = -\frac{(U_F - U_I)_{sys} - w}{T_{therm}} \qquad (5.60)$$

The global entropy change is:

$$(S_F - S_I)_{global} = (S_F - S_I)_{sys} + (S_F - S_I)_{therm}$$
$$= (S_F - S_I)_{sys} - \frac{(U_F - U_I)_{sys} - w}{T_{therm}} \qquad (5.61)$$

Removing for simplicity the subscript "sys":

$$w = U_F - U_I - T_{therm}(S_F - S_I) + T_{therm}(S_F - S_I)_{global} \qquad (5.62)$$

Using of the definition of the Helmholtz energy 5.2, to transform Eq. 5.62, we obtain:

$$A_I = U_I - T_I S_I \quad A_F = U_F - T_F S_F$$
$$\Downarrow$$
$$w = [A_F - (T_{therm} - T_F) S_F] - [A_I - (T_{therm} - T_I) S_I] \qquad (5.63)$$
$$+ T_{therm}(S_F - S_I)_{global}$$

The global entropy change is zero for a reversible process and positive for an irreversible one, we can conclude that:

$$w \geq [A_F - (T_{therm} - T_F) S_F] - [A_I - (T_{therm} - T_I) S_I] \qquad (5.64)$$

For a **monothermal** process where the initial and final temperatures of the system are equal to T_{therm}, we have:

$$T_{therm} = T_I = T_F$$
$$\Downarrow$$
$$w = (A_F - A_I) + T_{therm}(S_F - S_I)_{global} \qquad (5.65)$$
$$w \geq (A_F - A_I)$$

5 Auxiliary Functions : Enthalpy, Gibbs Energy, Helmholtz Energy

If $w > 0$, the change in the Helmholtz energy of the system corresponds to the minimum amount of (positive) work that must be done on the system to achieve the change. If $w < 0$, we can rewrite the last inequality 5.65 as :

$$-w \leq -(A_F - A_I) \tag{5.66}$$

The change in the Helmholtz energy of the system corresponds to the maximum amount of work *that can be obtained from the system.*

Example

A cylinder contains 0.1 mol of an ideal gas at 10 bar and 300 K. An isothermal reversible expansion brings its pressure to 1 bar. During the process, the gas temperature remains constant and its internal energy does not change. The change in the Helmholtz energy is (we use Eq. 4.22) :

$$A_F - A_I = U_F - U_I - T(S_F - S_I) = -T\left[-nR\ln\left(\frac{p_F}{p_I}\right)\right] = 300 \cdot 0.1 \cdot 8.3145 \ln 0.1 = -575 \text{ J}$$

The gas does work on the surroundings. The expression obtained for the change in A is identical to the work done on the system, taking into account the ideal gas law.

$$w_{sys} = -\int_{V_I}^{V_F} p_{ext} \, dV = -\int_{V_I}^{V_F} \frac{nRT}{V} dV = -nRT\ln\left(\frac{V_F}{V_I}\right) = nRT\ln\left(\frac{p_F}{p_I}\right) = -575 \text{ J}$$

Now, let us envisage the irreversible expansion from the same initial to the same final state during which the external pressure remains constant and equal to 1 bar. The work done on the system is :

$$w_{sys} = -p_{ext}(V_F - V_I) = -nRT\left(1 - \frac{p_{ext}}{p_I}\right) = -0.1 \cdot 8.3145 \cdot 300 \,(1 - 0.1) = -224 \text{ J}$$

As expected, the system provides less work during the irreversible process. Using Eq. 5.65 we find :

$$T_{therm}(S_F - S_I)_{global} = w_{sys} - (A_F - A_I) = -224 - (-575) = 351 \text{ J}$$

This represents the (absolute value of the) amount of work "wasted" due to the irreversibility.

5.7.2 Differential Form

For an *infinitesimal isothermal change*, we obtain :

$$\left.\begin{array}{l} dw = dA + T\, dS_{global} \\ dw \geq dA \end{array}\right\} \tag{5.67}$$

5.7.3 Gibbs Energy (Gibbs Function, Free Enthalpy)

Consider a *closed system* in contact with a thermal reservoir at temperature T_{therm}, which together make up an adiabatic closed system. Using 5.62 and the expression 5.3 for the Gibbs energy we obtain :

$$\left.\begin{array}{l} G_I = U_I + p_I V_I - T_I S_I \quad \Downarrow \quad G_F = U_F + p_F V_F - T_F S_F \\ w = [G_F - p_F V_F - (T_{therm} - T_F) S_F] \\ \quad - [G_I - p_I V_I - (T_{therm} - T_I) S_I] \\ \quad + T_{therm} (S_F - S_I)_{global} \end{array}\right\} \quad (5.68)$$

We can conclude that :

$$\left.\begin{array}{l} w \geq [G_F - p_F V_F - (T_{therm} - T_F) S_F] \\ \quad - [G_I - p_I V_I - (T_{therm} - T_I) S_I] \end{array}\right. \quad (5.69)$$

For a **monothermal monobaric process** at pressure $p_{ext} = p_I = p_F$ where $T_{therm} = T_I = T_F$, Eq. 5.68 and 5.69 become :

$$\left.\begin{array}{l} w = (G_F - G_I) - p_{ext}(V_F - V_I) + T_{therm}(S_F - S_I)_{global} \\ w \geq (G_F - G_I) - p_{ext}(V_F - V_I) \end{array}\right\} \quad (5.70)$$

The pressure containing term corresponds to the work done on the system due to its volume change. Other forms of work are written as w_{other}.

$$\left.\begin{array}{l} w_{other} = w - w_{volume} = w + p_{ext}(V_F - V_I) \\ \quad = (G_F - G_I) + T_{therm}(S_F - S_I)_{global} \\ \quad \Downarrow \\ w_{other} \geq (G_F - G_I) \end{array}\right\} \quad (5.71)$$

If $w_{other} > 0$, the change of the Gibbs energy (free enthalpy) is the minimum amount of work other than work due to volume change that has to be done on the system to achieve the change.
If $w_{other} < 0$ we can rewrite the last inequality of 5.71 as :

$$- w_{other} \leq -(G_F - G_I) \quad (5.72)$$

The change of the Gibbs energy (free enthalpy) corresponds to the maximum amount of work, other than work due to volume change, that can be obtained from the system.

5.7.4 Differential Form

For an *infinitesimal isothermal isobaric process*, we have :

$$\left.\begin{array}{l} dw_{other} = dG + T dS_{global} \\ dw_{other} \geq dG \end{array}\right\} \quad (5.73)$$

5.7.5 Spontaneous Evolution and Equilibrium Condition

Consider a *closed system in contact with a thermal reservoir* and in thermal equilibrium with it at the beginning and the end of the process. Assume no work can be done on the system ($w = 0$). Using 5.65, we have:

$$(A_F - A_I) = - T_{therm} (S_F - S_I)_{global} \leq 0 \quad (5.74)$$

A spontaneous evolution of a closed monothermal *system*, which cannot exchange work under any form with its surroundings takes place in such a way that A decreases. The system is at equilibrium when A has reached its minimum value.

Now, we envisage a closed system undergoing a monothermal and monobaric change and that can only receive work due to its volume change ($w_{other} = 0$). Using 5.71, we have:

$$(G_F - G_I) = - T_{therm} (S_F - S_I)_{global} \leq 0 \quad (5.75)$$

A spontaneous evolution of a closed monothermal monobaric *system* exchanging eventually only work due to its volume change will always take place in such a way that G decreases. The system is at equilibrium when G has reached its minimum value.

Example

Consider a cylinder closed by a piston in contact with a thermal reservoir at T. On one side of the piston, we have 0.1 mole of an ideal gas at a pressure of 10 bar. On the other side there is vacuum. A device limits the volume of the system to $V_{max} = 1$ l. The piston will naturally go to the right until the volume is V_{max}. During the process, no work is done on the system and since the temperature of gas is the same in the initial and final states, the system receives no heat.

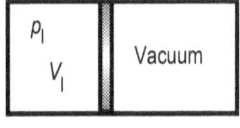

$$q_{sys} = - q_{therm} = 0$$

The global entropy change is the entropy change of the gas. We can represent the change of A as a function of the final volume V_F. We have:

$$A_F - A_I = - T(S_F - S_I)$$
$$= nRT \ln\left(\frac{p_F}{p_I}\right) = - nRT \ln\left(\frac{V_F}{V_I}\right)$$

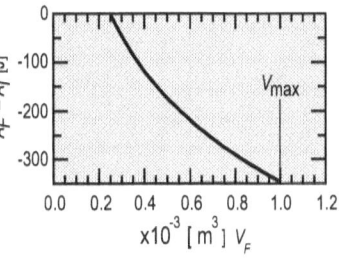

We find that the minimum of the Helmholtz energy is obtained when the piston is at the end of the cylinder.

6. Mixtures and Pure Substances : Partial Molar Quantities and Molar Quantities

6.1 Homogeneous Functions and their Properties

A function of several variables $F(x_1,..., x_i,...)$ is a homogeneous function of degree n of these variables if, for any positive value of λ, it satisfies :

$$F(\lambda x_1,..., \lambda x_i,...) = \lambda^n F(x_1,..., x_i,...) \tag{6.1}$$

By taking the derivative of both sides of Eq. 6.1 with respect to any of the x_i, we obtain :

$$\left. \begin{array}{l} \dfrac{\partial F(\lambda x_1,..., \lambda x_i,...)}{\partial(\lambda x_i)} \dfrac{\partial(\lambda x_i)}{\partial x_i} = \lambda^n \dfrac{\partial F(x_1,..., x_i,...)}{\partial x_i} \\[2mm] \dfrac{\partial(\lambda x_i)}{\partial x_i} = \lambda \Rightarrow \dfrac{\partial F(\lambda x_1,..., \lambda x_i,...)}{\partial(\lambda x_i)} = \lambda^{n-1} \dfrac{\partial F(x_1,..., x_i,...)}{\partial x_i} \end{array} \right\} \tag{6.2}$$

The partial derivatives of a homogeneous function of degree n with respect to one of the variables are therefore homogeneous functions of degree $n-1$.

Taking the derivative of both sides of 6.1 with respect to λ, then giving λ the value 1, we obtain :

$$\left. \begin{array}{l} \displaystyle\sum_i x_i \left[\dfrac{\partial F(\lambda x_1,..., \lambda x_i,...)}{\partial(\lambda x_i)} \right]_{\lambda x_j \ne i} = n\, \lambda^{n-1}\, F(x_1,..., x_i,...) \\[2mm] \text{with } \lambda = 1 \quad \Downarrow \\[2mm] \displaystyle\sum_i x_i \left[\dfrac{\partial F(x_1,..., x_i,...)}{\partial x_i} \right]_{x_j \ne i} = n\, F(x_1,..., x_i,...) \end{array} \right\} \tag{6.3}$$

The last relation of 6.3 is known as **Euler's identity**.

Example

Consider the function :

$$F(x_1,x_2) = 2 x_1^2 x_2^3 + 3 \frac{x_1^7}{x_2^2} \Rightarrow F(\lambda x_1, \lambda x_2) = 2 \lambda^2 x_1^2 \lambda^3 x_2^3 + 3 \frac{\lambda^7 x_1^7}{\lambda^2 x_2^2} = \lambda^5 F(x_1,x_2)$$

It is therefore a homogeneous function of degree 5 of the variables x_1, x_2. We can also verify Euler's identity. We have :

$$x_1 \frac{\partial F(x_1,x_2)}{\partial x_1} + x_2 \frac{\partial F(x_1,x_2)}{\partial x_2} = x_1 \left(4 x_1 x_2^3 + 21 \frac{x_1^6}{x_2^2} \right) + x_2 \left(6 x_1^2 x_2^2 - 6 \frac{x_1^7}{x_2^3} \right) = 5 F(x_1,x_2)$$

6.2 Extensive Variables

We consider a system that contains several species and where only one single homogeneous phase is present.

Experimental evidence teaches us that any extensive property is a homogeneous function of degree 1 of some of the other extensive variables of the system.

For U, H, A and G, as well as for an extensive property X, in the case where we consider X as $X(p, T,..., n_i,...)$:

$$\left. \begin{array}{l} U(\lambda V, \lambda S,..., \lambda n_i,...) = \lambda\, U(V,S,..., n_i,...) \\ H(\lambda S, p,..., \lambda n_i,...) = \lambda\, H(S, p,..., n_i,...) \\ A(\lambda V, T,..., \lambda n_i,...) = \lambda\, A(V, T,..., n_i,...) \\ G(T, p,..., \lambda n_i,...) = \lambda\, G(T, p,..., n_i,...) \\ X(T, p,..., \lambda n_i,...) = \lambda\, X(T, p,..., n_i,...) \end{array} \right\} \quad (6.4)$$

6.3 Intensive Variables

A first *order partial derivative* of any extensive variable with respect to another extensive variable is an *intensive variable*. We indicate by a * superscript the variables relative to the system obtained after multiplying the extensive variables by a factor λ. We have :

$$\left. \begin{array}{l} V^* = \lambda V \\ S^* = \lambda S \\ ..., n_i^* = \lambda n_i,... \end{array} \right\} \Rightarrow \frac{\partial V^*}{\partial V} = \frac{\partial S^*}{\partial S} = ... = \frac{\partial n_i^*}{\partial n_i} = ... = \lambda \quad (6.5)$$

We can evaluate the partial derivatives of U with respect to S, for each side of Eq. 6.4 using relations 6.5 :

6 Mixtures and Pure Substances : Partial Molar Quantities and Molar Quantities

$$\frac{\partial}{\partial S}\left[U(V^*,S^*,n_i^*...)\right]_{V^*,n_i^*} = \left[\frac{\partial}{\partial S^*}U(V^*,S^*,...,n_i^*,...)\right]_{V^*,n_i^*}\frac{\partial S^*}{\partial S}$$

$$= \left[\frac{\partial}{\partial S^*}U(V^*,S^*,...,n_i^*,...)\right]_{V^*,n_i^*}\lambda$$

(6.6)

and for the right hand side of Eq. 6.4 :

$$\frac{\partial}{\partial S}\left[\lambda\, U(V,S,...,n_i,...)\right]_{V,n_i} = \lambda\,\frac{\partial}{\partial S}\left[U(V,S,...,n_i,...)\right]_{V,n_i} \quad (6.7)$$

We find :

$$\left[\frac{\partial}{\partial S^*}U(V^*,S^*,...,n_i^*,...)\right]_{V^*,n_i^*} = \left[\frac{\partial}{\partial S}U(V,S,...,n_i,...)\right]_{V,n_i} = T$$

(6.8)

Similarly :

$$\left(\frac{\partial U^*}{\partial V^*}\right)_{S^*,n_i^*} = \left(\frac{\partial U}{\partial V}\right)_{S,n_i} = -p \quad (6.9)$$

$$\left(\frac{\partial U^*}{\partial n_i^*}\right)_{S^*,V^*,n_j^*} = \left(\frac{\partial U}{\partial n_i}\right)_{S,V,n_j} = \mu_i \quad (6.10)$$

Intensive variables do not change when extensive variables are multiplied by a real positive integer. This result shows that *chemical potentials* are *intensive variables*. Mole fractions are another example of intensive variables. We have :

$$x_i = \frac{n_i}{\sum_i n_i} = \frac{\lambda n_i}{\sum_i \lambda n_i} \quad (6.11)$$

The chemical potential are homogeneous functions of degree 0 of the n_i. We have using *Euler's identity* for a homogeneous function of degree 0 :

$$\left.\begin{array}{l} \mu_i(T,p,...,\lambda n_j,...) = \mu_i(T,p,...,n_j,...) \\[6pt] \displaystyle\sum_j n_j\left(\frac{\partial \mu_i}{\partial n_j}\right)_{T,p,n_k\neq n_j} = 0 \end{array}\right\} \quad (6.12)$$

6.4 Explicit Expressions for Various Extensive Variables

Using Euler's identity, Eq. 5.21 and 5.23, we find the explicit expression for the corresponding extensive variables:

$$\left. \begin{array}{c} U = \left(\dfrac{\partial U}{\partial V}\right)_{S, n_i} V + \left(\dfrac{\partial U}{\partial S}\right)_{V, n_i} S + \sum_i \left(\dfrac{\partial U}{\partial n_i}\right)_{S, V, n_j} n_i \\ \\ = -pV + TS + \sum_i \mu_i n_i \end{array} \right\} \quad (6.13)$$

From this result, we get:

$$H = U + pV = TS + \sum_i \mu_i n_i \quad (6.14)$$

For A and G:

$$A = U - TS = -pV + \sum_i \mu_i n_i \quad (6.15)$$

$$G = U + pV - TS = \sum_i \mu_i n_i \quad (6.16)$$

Example

The chemical potential of a gas in an ideal gas mixture is:

$$\mu_i = \mu_i^\ominus(T) + RT \ln p + RT \ln x_i$$

where p is the system pressure and x_i is the mole fraction of species i in the mixture. The quantity $\mu_i^\ominus(T)$ is the standard chemical potential which we will define later (chapter 7) that corresponds to $x_i = 1$ and $p = 1$ bar. We can thus express the Gibbs energy of the mixture as:

$$G = \sum_i n_i \left(\mu_i^\ominus(T) + RT \ln p + RT \ln x_i \right)$$

6.5 Gibbs-Duhem Equation

The differential of G can also be obtained from Eq. 6.16. We have (5.23):

6 Mixtures and Pure Substances : Partial Molar Quantities and Molar Quantities

$$\left. \begin{array}{l} dG = V\,dp - S\,dT + \sum_i \mu_i\,dn_i \\[6pt] dG = \sum_i n_i\,d\mu_i + \sum_i \mu_i\,dn_i \end{array} \right\} \quad (6.17)$$

The two expressions must be equal and we must have :

$$- V\,dp + S\,dT + \sum_i n_i\,d\mu_i = 0 \quad (6.18)$$

The variables p, T and the μ_i are therefore not independent. Equation 6.18 is known as the **Gibbs-Duhem equation**.

6.6 Partial Molar Quantities

For any extensive variable considered as a function of T and p, the only extensive variables needed to define the state of the system are the number of moles of the various species. Using Euler's identity for a homogeneous function of degree 1, we have :

$$\left. \begin{array}{l} X = \sum_i n_i \left(\dfrac{\partial X}{\partial n_i} \right)_{T,\,p,\,n_j} = \sum_i n_i\,\overline{X_i} \\[10pt] \text{with } \overline{X_i} = \left(\dfrac{\partial X}{\partial n_i} \right)_{T,\,p,\,n_j} \end{array} \right\} \quad (6.19)$$

$\overline{X_i}$ is the **partial molar quantity** of species i corresponding to the extensive property X. It is an *intensive variable*. In general, *it depends on the composition of the system*. We can write the corresponding equations for any extensive variable.

$$\left. \begin{array}{ll} U = \sum_i n_i\,\overline{U_i} & H = \sum_i n_i\,\overline{H_i} \\[6pt] A = \sum_i n_i\,\overline{A_i} & V = \sum_i n_i\,\overline{V_i} \\[6pt] S = \sum_i n_i\,\overline{S_i} & G = \sum_i n_i\,\overline{G_i} = \sum_i n_i\,\mu_i \\[6pt] C_p = \sum_i n_i\,\overline{C_{pi}} & C_V = \sum_i n_i\,\overline{C_{Vi}} \end{array} \right\} \quad (6.20)$$

where, for example:

$$\overline{U_i} = \left(\frac{\partial U}{\partial n_i}\right)_{T, p, n_j} \quad \overline{V_i} = \left(\frac{\partial V}{\partial n_i}\right)_{T, p, n_j} \quad \overline{G_i} = \left(\frac{\partial G}{\partial n_i}\right)_{T, p, n_j} = \mu_i$$

(6.21)

The variables T and p are kept constant in the evaluation of the derivatives in Eq. 6.21. In general, partial molar quantities depend on the composition of the system.

Example

The enthalpy of n_i moles of a gas in an ideal gas mixture is given by:

$$H_i = n_i \left[H_{m\,i}(T_0) + C_{p,\,m\,i}(T - T_0) \right]$$

assuming the heat capacity at constant pressure is a constant.

The expression of the partial molar enthalpy is:

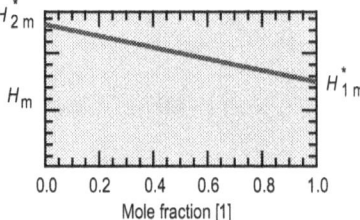

$$\overline{H_i} = \left(\frac{\partial H_i}{\partial n_i}\right)_{T, p} = H_{m\,i}(T_0) + C_{p,\,m\,i}(T - T_0)$$

In this case, the partial molar enthalpy turns out to be independent of the composition. The enthalpy of a mixture of two gases forming an ideal gas mixture is:

$$H = n_1 \overline{H_1} + n_2 \overline{H_2}$$

$$= n_1 \left(H_{m\,1}(T_0) + C_{p,\,m\,1}(T - T_0) \right) + n_2 \left(H_{m\,2}(T_0) + C_{p,\,m\,2}(T - T_0) \right)$$

where subscripts 1 and 2 refer to each species. At a given temperature, the graph of the enthalpy of one mole ($n_1 = x_1$, $n_2 = 1 - x_1$) of such a mixture as a function of composition is a straight line.

We can write two expressions for the differential of X. We find:

$$\left.\begin{aligned} dX &= \left(\frac{\partial X}{\partial p}\right)_{T, n_i} dp + \left(\frac{\partial X}{\partial T}\right)_{p, n_i} dT + \sum_i \left(\frac{\partial X}{\partial n_i}\right)_{p, T, n_j} dn_i \\ dX &= \sum_i n_i\, d\overline{X_i} + \sum_i \overline{X_i}\, dn_i \end{aligned}\right\}$$

(6.22)

We obtain a relation between intensive variables, analogous to the Gibbs-Duhem equation.

$$\sum_i n_i\, d\overline{X_i} - \left(\frac{\partial X}{\partial p}\right)_{T, n_i} dp - \left(\frac{\partial X}{\partial T}\right)_{p, n_i} dT = 0 \qquad (6.23)$$

6.7 Molar Quantities. Pure Substances

Equations 6.20 can be applied to a pure substance present in only one phase:

$$\left.\begin{array}{ll} U = n\,\overline{U} = n\,U_m & S = n\,\overline{S} = n\,S_m \\ H = n\,\overline{H} = n\,H_m & G = n\,\overline{G} = n\mu = n\,G_m \\ A = n\,\overline{A} = n\,A_m & C_p = n\,\overline{C_p} = n\,C_{p,m} \\ V = n\,\overline{V} = n\,V_m & C_V = n\,\overline{C_V} = n\,C_{V,m} \end{array}\right\} \quad (6.24)$$

For pure species, the composition is fixed. The corresponding quantities are called **molar quantities** noted by a subscript m.

Example

The molar entropy of a pure ideal gas at temperature T and pressure p is given by (See chapter 7):

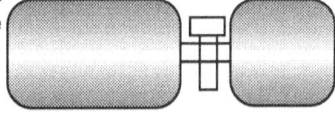

$$S_m = S^\ominus(T) - R \ln p$$

where $S^\ominus(T)$ is the standard molar entropy of the gas at T. We consider a two compartment vessel in contact with a thermal reservoir at 300 K. Compartment 1 contains 0.1 mol of a gas at pressure 10^5 Pa. Compartment 2 contains 0.2 mol of the *same gas* at 10^6 Pa. The stopcock separating the compartments is open. The equilibrium pressure is obtained using the ideal gas law:

$$RT = \frac{p_{eq}(V_1 + V_2)}{n_1 + n_2} = \frac{p_{eq}\left(\dfrac{n_1 RT}{p_1} + \dfrac{n_2 RT}{p_2}\right)}{n_1 + n_2} \Rightarrow p_{eq} = \frac{n_1 + n_2}{\dfrac{n_1}{p_1} + \dfrac{n_2}{p_2}}$$

Finally, we obtain for the equilibrium pressure the simple expression:

$$p_{eq} = \frac{(n_1 + n_2) p_1 p_2}{n_1 p_2 + n_2 p_1} = \frac{0.3 \cdot 10^5 \cdot 10^6}{0.1 \cdot 10^6 + 0.2 \cdot 10^5} = 2.5\ 10^5 \text{ Pa} = 2.5 \text{ bar}$$

The change of the entropy of the gas for this process is:

$$S_F - S_I = -(n_1 + n_2) R \ln p_{eq} + (n_1 R \ln p_1 + n_2 R \ln p_2) \simeq 1.54 \text{ J K}^{-1}$$

6.8 Other Relations

We can write using 5.30 and 5.31:

$$\left.\begin{array}{l} \left(\dfrac{\partial \mu_i}{\partial T}\right)_{p,\,n_i,\,n_j} = -\left(\dfrac{\partial S}{\partial n_i}\right)_{T,\,p,\,n_j} = -\overline{S_i} \\[1em] \left(\dfrac{\partial \mu_i}{\partial p}\right)_{T,\,n_i,\,n_j} = \left(\dfrac{\partial V}{\partial n_i}\right)_{T,\,p,\,n_j} = \overline{V_i} \end{array}\right\} \quad (6.25)$$

Example

The chemical potential of a gas in an ideal gas mixture is :

$$\mu_i = \mu_i^\ominus(T) + RT \ln p + RT \ln x_i$$

where p is the system pressure and x_i is the mole fraction of species i in the mixture. The quantity $\mu_i^\ominus(T)$ is the standard chemical potential which we will define later (chapter 7) that corresponds to $x_i = 1$ and $p = 1$ bar. We have :

$$\overline{V}_i = \left(\frac{\partial \mu_i}{\partial p}\right)_{T, n_i, n_j} = \frac{RT}{p} = V_{mi}$$

The partial molar volume of an ideal gas in an ideal gas mixture is independent of composition and equal to its molar volume.

Taking the partial derivative with respect to n_i of Eq. 5.1 to 5.3, we get the following relations between intensive variables :

$$\left.\begin{array}{l}\left(\dfrac{\partial H}{\partial n_i}\right)_{T, p, n_j} = \left(\dfrac{\partial U}{\partial n_i}\right)_{T, p, n_j} + p\left(\dfrac{\partial V}{\partial n_i}\right)_{T, p, n_j} \\ \overline{H}_i \quad\quad = \overline{U}_i \quad\quad + p\,\overline{V}_i \end{array}\right\} \quad (6.26)$$

$$\left.\begin{array}{l}\left(\dfrac{\partial A}{\partial n_i}\right)_{T, p, n_j} = \left(\dfrac{\partial U}{\partial n_i}\right)_{T, p, n_j} - T\left(\dfrac{\partial S}{\partial n_i}\right)_{T, p, n_j} \\ \overline{A}_i \quad\quad = \overline{U}_i \quad\quad - T\,\overline{S}_i \end{array}\right\} \quad (6.27)$$

$$\left.\begin{array}{l}\left(\dfrac{\partial G}{\partial n_i}\right)_{T, p, n_j} = \left(\dfrac{\partial H}{\partial n_i}\right)_{T, p, n_j} - T\left(\dfrac{\partial S}{\partial n_i}\right)_{T, p, n_j} \\ \overline{G}_i = \mu_i \quad = \overline{H}_i \quad\quad - T\,\overline{S}_i \end{array}\right\} \quad (6.28)$$

we also have (Eq. 6.25) :

$$\left.\begin{array}{c} \overline{G}_i = \mu_i = \overline{H}_i - T\,\overline{S}_i = \overline{H}_i + T\left(\dfrac{\partial \mu_i}{\partial T}\right)_{p, n_i, n_j} \\ \Downarrow \\ \left[\dfrac{\partial}{\partial T}\left(\dfrac{\mu_i}{T}\right)\right]_{p, n_i, n_j} = \dfrac{1}{T}\left(\dfrac{\partial \mu_i}{\partial T}\right)_{p, n_i, n_j} - \dfrac{\mu_i}{T^2} = -\dfrac{\overline{H}_i}{T^2} \end{array}\right\} \quad (6.29)$$

6 Mixtures and Pure Substances : Partial Molar Quantities and Molar Quantities

For a pure substance, we get :

$$\left(\frac{\partial \mu}{\partial p}\right)_T = \overline{V} = V_m \qquad \left(\frac{\partial \mu}{\partial T}\right)_p = -\overline{S} = -S_m$$

$$\mu = H_m - T S_m = H_m + T\left(\frac{\partial \mu}{\partial T}\right)_p \qquad (6.30)$$

$$\left[\frac{\partial}{\partial T}\left(\frac{\mu}{T}\right)\right]_p = \frac{1}{T}\left(\frac{\partial \mu}{\partial T}\right)_p - \frac{\mu}{T^2} = -\frac{H}{T^2} = -\frac{H_m}{T^2}$$

Example

It is of interest to mention here the principle of the measurement of partial molar quantities. Let us assume that at p and T, the volume of a mixture (of liquids) has been measured as a function of composition. We represent the volume per mole of mixture as a function of the composition of the mixture in species 1. The equation of the curve representing the molar volume as a function of composition is :

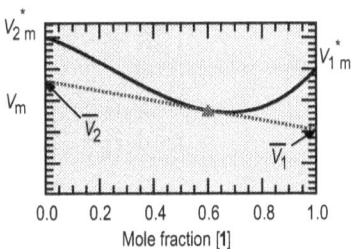

$$V_m = \frac{V}{n_1 + n_2} = \frac{n_1 \overline{V_1} + n_2 \overline{V_2}}{n_1 + n_2} = x_1 \overline{V_1} + x_2 \overline{V_2} = x_1 \overline{V_1} + (1 - x_1) \overline{V_2}$$

We have using the expression of V_m and 6.23 at T and p constant :

$$dV_m = (\overline{V_1} - \overline{V_2}) dx_1 + x_1 d\overline{V_1} + x_2 d\overline{V_2}$$
$$\Downarrow \quad \text{with} \quad n_1 d\overline{V_1} + n_2 d\overline{V_2} = 0$$
$$dV_m = (\overline{V_1} - \overline{V_2}) dx_1$$

So that the slope of the tangent to the curve at a given composition (for example $x_1 = x_0 = 0.6$) is :

$$\left.\frac{dV_m}{dx_1}\right|_{x_1 = x_0} = \overline{V_1}(x_0) - \overline{V_2}(x_0)$$

the difference of the partial molar volumes at that concentration. The equation of the tangent to the curve at composition x_0 is :

$$y = \left(\overline{V_1}(x_0) - \overline{V_2}(x_0)\right) x_1 + \overline{V_2}(x_0)$$

Taking into account the fact that the tangent passes through the point $y = V_m(x_0)$. The ordinate of the tangent at $x_1 = 0$ and $x_1 = 1$ corresponds to the values of $\overline{V_2}(x_0)$ and $\overline{V_1}(x_0)$ respectively. In practice, the effects are a lot less pronounced than shown on the graph. More sensitive graphs can be drawn.

7. Thermodynamics of Gases

7.1 Pure Ideal Gas
7.1.1 Chemical Potential of a Pure Ideal Gas
We use the equation of state for n moles:
$$pV = nRT \tag{1.16}$$
The differential of the Gibbs energy is (5.22):
$$dG = -SdT + Vdp + \mu\, dn \tag{7.1}$$
Using Schwarz theorem (see also 6.30):
$$\left(\frac{\partial \mu}{\partial p}\right)_{T,n} = \left(\frac{\partial V}{\partial n}\right)_{T,p} = V_m \quad \Rightarrow \quad \left(\frac{\partial \mu}{\partial p}\right)_{T,n} = \frac{RT}{p} \tag{7.2}$$
The integration of this partial differential equation gives us the dependence of the chemical potential of a pure ideal gas on pressure:
$$\int_{\mu^0}^{\mu} d\mu = \int_{p^0}^{p} RT \frac{dp}{p} \quad \Rightarrow \quad \mu = \mu^0(T) + RT \ln\left(\frac{p}{p^0}\right) \tag{7.3}$$
The chemical potential of a pure ideal gas has the value $\mu^0(T)$ at pressure p^0. $\mu^0(T)$ is a function of temperature only.

7.1.2 Selection of the Standard State Pressure
The special superscript $^\ominus$ is used to refer to a standard state quantity. The chemical potential of an ideal gas is written:
$$\mu = \mu^\ominus(T) + RT \ln \frac{p}{p^\ominus} = \mu^\ominus(T) + RT \ln p \tag{7.4}$$
where $\mu^\ominus(T)$ is the chemical potential of the ideal gas under consideration at standard state pressure, $p^\ominus$. $\mu^\ominus(T)$ is known as the **standard chemical potential** at temperature T.
The right hand side of Eq. 7.4 is valid because the standard state pressure always has a value of 1 (either 1 atm or 1 bar according to the selected convention) (1 atm = 1.01325 10^5 Pa). The most recent convention for the standard state pressure is a pressure of 1 bar (1 bar = 10^5 Pa = 750.0062 mmHg). In view of this choice, gas pressures should be expressed in bar in equations involving chemical potentials.

Example

Let us calculate the change in the chemical potential of an ideal gas when its pressure varies from 1 bar to 2 bar at 300 K. We have:

$$\mu(T, 2) - \mu(T, 1) = 8.3145 \cdot 300 \cdot \ln\left(\frac{2}{1}\right) \simeq 1729 \text{ J mol}^{-1}$$

At a temperature T, the difference of the standard chemical potentials using 1 bar or 1 atm as the standard pressure is:

$$\mu_b^\ominus - \mu_a^\ominus = RT \ln\left(\frac{p_b^\ominus}{p_a^\ominus}\right) = -T\delta \quad \text{where } \delta = R \ln 1.01325 = 0.10944 \text{ J mol}^{-1} \text{ K}^{-1}$$

At 298.15 K the difference of the two standard potentials is 32 J mol^{-1}.

7.1.3 Mathematical Expressions of other Thermodynamic Functions of Ideal Gases

The molar entropy of an ideal gas is given by (Eq. 5.30):

$$\left(\frac{\partial \mu}{\partial T}\right)_{p,n} = -\left(\frac{\partial S}{\partial n}\right)_{T,p} = -S_m \Rightarrow S_m = -\frac{d\mu^\ominus}{dT} - R \ln p = S^\ominus(T) - R \ln p$$

(7.5)

where $S^\ominus(T)$ is the **standard molar entropy** of the ideal gas considered, at temperature T.
Since $\mu^\ominus$ only depends on T, $S^\ominus$ depends only on T as well. Various expressions of the molar enthalpy are:

$$G = H - TS \Rightarrow \begin{cases} H_m = \mu + TS_m \\ = \mu^\ominus + RT \ln p + T\left(-\dfrac{d\mu^\ominus}{dT} - R \ln p\right) \\ = \mu^\ominus - T\dfrac{d\mu^\ominus}{dT} = -T^2 \dfrac{d\left(\dfrac{\mu^\ominus}{T}\right)}{dT} = H^\ominus \end{cases}$$

(7.6)

H_m is identical to the standard molar enthalpy, $H^\ominus$, independent of pressure.

7.2 Mixtures of Ideal Gases

7.2.1 Basic Properties. Ideal Gas Mixture. Dalton's Law

A mixture of ideal gases is an **ideal gas mixture** if both of the following properties are valid:

- The internal energy of the mixture at temperature T is the sum of the internal energy of the species (ideal gases) of the mixture taken separately at the same temperature T.

7 Thermodynamics of Gases

$$U = \sum_i U_i \quad (7.7)$$

- The enthalpy of the mixture at temperature T is the sum of the enthalpy of the species (ideal gases) of the mixture taken separately at the same temperature T.

$$H = \sum_i H_i \quad (7.8)$$

The equation of state of a pure ideal gas and the definition of H allow us to write:

$$H = U + pV = \sum_i (U_i + p_i V_i) = U + \sum_i n_i RT = U + nRT \quad (7.9)$$

The number of moles n identifies with the sum of the number of moles of the species as it should.

Example

Consider an ideal gas mixture containing two species. Assume that the heat capacities at constant volume are constant. The change of the internal energy of the mixture, when the temperature changes, is:

$$U_2 - U_1 = (n_1 C_{V,m\,1} + n_2 C_{V,m\,2})(T_2 - T_1)$$

unaffected by volume.
The change of the enthalpy is, using 5.52:

$$H_2 - H_1 = (n_1 C_{V,m\,1} + n_2 C_{V,m\,2})(T_2 - T_1) + (n_1 + n_2) R (T_2 - T_1)$$
$$= (n_1 C_{p,m\,1} + n_2 C_{p,m\,2})(T_2 - T_1)$$

unaffected by pressure.

If the n_i moles of species i occupied by themselves the entire volume, the pressure as given by the equation of state would be:

$$p_i V = n_i RT \quad (7.10)$$

The pressure p_i is known as the partial pressure of species i in the mixture. We can write using the fact that there are n moles of gas:

$$\frac{RT}{V} = \frac{p_1}{n_1} = \ldots = \frac{p_i}{n_i} = \ldots = \frac{\sum_i p_i}{\sum_i n_i} = \frac{p}{n} \quad \text{using } pV = nRT \quad (7.11)$$

$$p = \sum_i p_i \qquad p_i = p\,x_i \qquad x_i = \frac{n_i}{n} \quad (7.12)$$

The pressure of the mixture is equal to the sum of the partial pressures of the species (**Dalton's law**). We also obtained the expression of the partial pressure of species i as a function of its mole fraction, x_i, in the mixture and of the total pressure.

Example
In a $3 \cdot 10^{-3}$ m^3 vessel at 400 K, we have 0.4 mol of N_2 and 0.1 mol of O_2 assumed to behave ideally. Let us find the partial pressure of each gas and the total pressure. We have:

$$p_{N_2} = \frac{0.4 \cdot 8.3145 \cdot 400}{3 \cdot 10^{-3}} \simeq 4.434 \cdot 10^6 \text{ Pa} = 44.34 \text{ bar}$$

$$p_{O_2} = \frac{0.1 \cdot 8.3145 \cdot 400}{3 \cdot 10^{-3}} \simeq 1.109 \cdot 10^6 \text{ Pa} = 11.09 \text{ bar}$$

the total pressure is the sum of the partial pressure:

$$p_{total} = \frac{0.5 \cdot 8.3145 \cdot 400}{3 \cdot 10^{-3}} \simeq 5.543 \cdot 10^6 \text{ Pa} = 55.43 \text{ bar}$$

We can obtain the partial molar volume using:

$$V = \frac{nRT}{p} = \frac{RT}{p} \sum_i n_i \quad \Rightarrow \quad \left(\frac{\partial V}{\partial n_i}\right)_{T,p,n_j} = \overline{V_i} = \frac{RT}{p} = V_m \quad (7.13)$$

The partial molar volume of an ideal gas in an ideal gas mixture is the same as its molar volume and it is also the same for all gases at given p and T. As a consequence, the molar volume of a mixture in an ideal gas mixture is independent of the composition of the mixture.

7.2.2 Chemical Potential of an Ideal Gas in an Ideal Gas Mixture

We start from the differential expression of the Gibbs energy and take into account the ideal gas law to express V and the fact that the pressure is the sum of the partial pressures.

$$\left. \begin{array}{l} dG = V dp - S dT + \sum_i \mu_i dn_i \text{ with} \\ \\ = \sum_i \frac{n_i RT}{p_i} dp_i - S dT + \sum_i \mu_i dn_i \end{array} \right\} \begin{array}{l} V = \ldots = \frac{n_i RT}{p_i} = \ldots \\ \\ dp = \sum_i dp_i \end{array} \right\} \quad (7.14)$$

We can find the variation of μ_i with the partial pressures. Using Schwartz theorem, we find:

7 Thermodynamics of Gases

$$\left(\frac{\partial \mu_i}{\partial p_i}\right)_{p_j, T, n_i, n_j} = \frac{\partial}{\partial n_i}\left(\frac{n_i R T}{p_i}\right)_{p_i, p_j, T, n_j} = \frac{R T}{p_i}$$

$$\left(\frac{\partial \mu_i}{\partial p_j}\right)_{p_i, T, n_i, n_j} = \frac{\partial}{\partial n_i}\left(\frac{n_i R T}{p_j}\right)_{p_i, p_j, T, n_j} = 0$$

(7.15)

From Eq. 7.15 and 6.12, one can find the expression for the chemical potential of an ideal gas in a mixture.

$$\mu_i = \mu_i^\ominus(T) + R T \ln p_i = \mu_i^\ominus(T) + R T \ln \frac{p_i}{p^\ominus}$$

$$= \mu_i^\ominus(T) + R T \ln p + R T \ln x_i$$

(7.16)

From this expression, we can get expressions for other thermodynamic properties. Using 6.29, we get:

$$\overline{H}_i = -T^2\left[\frac{\partial}{\partial T}\left(\frac{\mu_i}{T}\right)\right]_{p, n_i, n_j} = -T^2\left[\frac{\partial}{\partial T}\left(\frac{\mu_i^\ominus}{T}\right)\right]_{p, n_i, n_j}$$

$$= -T^2 \frac{d\left(\frac{\mu_i^\ominus}{T}\right)}{dT} = H_i^\ominus = H_{i, m}$$

(7.17)

The partial molar entropy of an ideal gas in an ideal gas mixture depends on the composition of the mixture.

$$\overline{S}_i = -\left(\frac{\partial \mu_i}{\partial T}\right)_{p, n_i, n_j} = -\frac{d\mu_i^\ominus}{dT} - R \ln p - R \ln x_i$$

$$= S_i^\ominus(T) - R \ln p - R \ln x_i = S_{i, m}^* - R \ln x_i$$

(7.18)

$S_i^\ominus(T)$ is the standard molar entropy of species i. $S_{i, m}^*$ is the molar entropy of species i pure at pressure p. The superscript * is conventionally used to indicate a pure species property.

Example

Consider an ideal gas mixture assuming that the $C_{p, m i}$ are constant. From the enthalpy of the pure species, we obtain the partial molar enthalpy:

$$H_i = n_i\left[H_{m\, i}(T_0) + C_{p, m i}(T - T_0)\right]$$

$$\Downarrow$$

$$\overline{H}_i = \left(\frac{\partial H_i}{\partial n_i}\right)_{T, p} = H_{m\, i}(T_0) + C_{p, m i}(T - T_0) = H_i^\ominus(T)$$

The partial molar heat capacity is given by:

$$\overline{C}_{pi} = \frac{d\overline{H}_i}{dT} = \frac{dH_i^\ominus}{dT} = C_{pi}^\ominus = C_{p,mi}$$

We can also find how the standard molar entropy varies with temperature. We have:

$$T\frac{dS_i^\ominus(T)}{dT} = \overline{C}_{pi}^\ominus = C_{p,mi} \Rightarrow S_i^\ominus(T) = S_i^\ominus(T_0) + C_{p,mi}\ln\left(\frac{T}{T_0}\right)$$

7.2.3 Mixing Properties

In its initial state, a system is made of several pure ideal gases, *all at pressure p and temperature T*. We mix them reversibly to obtain a mixture at the same pressure p and temperature T. The entropy change for such a process is known as the *entropy of mixing*:

$$\Delta_{mix}S = \sum_i n_i \overline{S}_i - \sum_i n_i S_{i,m}^* = -\sum_i n_i R \ln x_i \qquad (7.19)$$

Since the mole fractions x_i are smaller than 1, the system entropy increases during the mixing process. We also have:

$$\left.\begin{array}{l} \Delta_{mix}H = H_F - H_I = 0 \\ \Downarrow \\ \Delta_{mix}G = G_F - G_I = \Delta_{mix}H - T\Delta_{mix}S \\ \qquad = T\sum_i n_i R \ln x_i \end{array}\right\} \qquad (7.20)$$

The enthalpy and internal energy of mixing are zero and the Helmholtz energy of mixing is given by:

$$\left.\begin{array}{l} \Delta_{mix}U = U_F - U_I = 0 \\ \Downarrow \\ \Delta_{mix}A = A_F - A_I = \Delta_{mix}U - T\Delta_{mix}S \\ \qquad = T\sum_i n_i R \ln x_i \end{array}\right\} \qquad (7.21)$$

Note that the change in A is equal to the work done on the system during this reversible process. Since the internal energy of the gases does not change (T is constant), the system also receives heat during this reversible process.

7.2.4 Irreversible Mixing of Two Ideal Gases

Consider two gases in two different compartments of a vessel in contact with a thermal reservoir. Both gases are at the same temperature T and pressure p. When mixed, they form an ideal gas mixture. The vessels are connected via a stopcock that is now open. The final pressure is p. No work is done on the system. Since the temperature of the system is constant, its internal energy does not change. The first law of thermodynamics therefore tells us that no heat exchange takes place. The entropy change of the thermal reservoir, ΔS_{therm}, is therefore zero. The state functions of the system undergo exactly the same changes as obtained for the reversible mixing. The global change of entropy (system plus thermal reservoir) is, in this case, the change of the entropy of the system, which is equal to the entropy of mixing and is positive. This indicates the irreversibility of such a mixing process.

Example

We consider a two compartment vessel in contact with a thermal reservoir at 300 K. One compartment contains 0.2 mol of O_2 and the other, 0.8 mol of N_2. The pressure is the same in both compartments and is 10^5 Pa. When the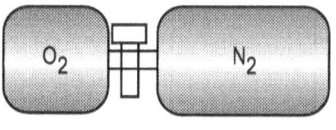
stopcock is open, the gases mix. We assume that the gases behave ideally. At equilibrium, the mixture has a uniform composition. The pressure remains constant since the total volume, the temperature and the number of moles are constant.

$$\Delta_{mix}S = -\sum_i n_i R \ln x_i = -8.3145 \, (0.2 \cdot \ln 0.2 + 0.8 \ln 0.8) = 4.16 \text{ J K}^{-1}$$

We find, as expected, that the entropy change of the global system is positive.

7.3 Pure Real Gases

7.3.1 Molecular Interactions in Real Gases

A real gas can behave as an ideal gas if its temperature is significantly above its temperature of liquefaction and its pressure is sufficiently low, practically of the order of atmospheric pressure. Gas molecules interact via forces known as "van der Waals forces". The potential energy of interaction of two non polar molecules can be described to a good approximation by the **Lennard-Jones** potential :

$$E_{interaction} = \frac{c_2}{r^{12}} - \frac{c_1}{r^6} \tag{7.22}$$

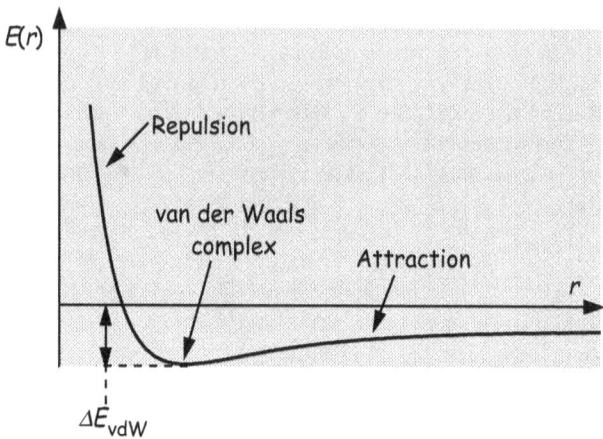

Figure 7.1 Schematic representation of the interaction energy between two molecules of gas as a function of the separation distance.

The constants C_1 and C_2 depend on the nature of the molecules. We have given a schematic representation of the interaction energy, $E(r)$, in Fig. 7.1.

When molecules are sufficiently apart, the attractive forces are the dominating interaction. At very short separation distances, the repulsion forces dominate and the potential energy becomes strongly positive. The distance where repulsive and attractive forces are equal corresponds to the van der Waals complex.

7.3.2 Chemical Potential of a Pure Real Gas

Since we observe that real gases behave as ideal gases at low enough pressure, we express the chemical potential of such a gas by a mathematical expression similar to that obtained for an ideal gas:

$$\left. \begin{array}{l} \mu = \mu^\ominus(T) + R\,T\ln f = \mu^\ominus(T) + R\,T\ln p + R\,T\ln \phi \\ \text{with} \quad f = p\,\phi \quad \phi \to 1 \text{ when } p \to 0 \end{array} \right\} \quad (7.23)$$

f is called the **fugacity** of the gas and has units of pressure. ϕ is the **fugacity coefficient**, is dimensionless and is a function of temperature and pressure.

The standard state for a real gas is the state that corresponds to the standard state pressure (1 bar or 1 atm according to the convention selected) and a value of the fugacity coefficient, $\phi = 1$. For many real gases at standard state pressure, the fugacity coefficient, ϕ, is quite close to 1. We can also write:

$$\mu_{real} = \mu_{ideal} + R\,T\ln \phi \qquad (7.24)$$

7 Thermodynamics of Gases

The fugacity coefficient represents the deviation of the chemical potential of the real gas compared to the corresponding ideal gas for given values of pressure and temperature.

7.3.3 Fugacity Coefficient of a Pure Real Gas

If we know the equation of state of the gas, the fugacity coefficient at some given temperature can be obtained in the following way:

$$\left(\frac{\partial \mu}{\partial p}\right)_T = V_m \implies V_m = \frac{RT}{p} + RT\left(\frac{\partial \ln \phi}{\partial p}\right)_T \quad (7.25)$$

$$\left(\frac{\partial \ln \phi}{\partial p}\right)_T = \frac{V_m}{RT} - \frac{1}{p} \quad (7.26)$$

$$\int_{\phi=1}^{\phi} d\ln \phi = \int_0^p \left(\frac{p V_m}{RT} - 1\right)\frac{dp}{p} = \int_0^p (Z-1)\frac{dp}{p} \quad (7.27)$$

where $Z = \dfrac{p V_m}{RT}$

The function Z is the **compressibility factor** (which can be expressed in terms of p using the equation of state).

$$\ln \phi = \int_0^p (Z-1)\frac{dp}{p} \quad (7.28)$$

7.3.4 The Virial Equation

The product $p V_m$ can be expressed as:

$$\frac{p V_m}{RT} = 1 + B'p + C'p^2 + D'p^3 + \ldots \quad (7.29)$$

where
- B' = second virial coefficient
- C' = third virial coefficient
- D' = fourth virial coefficient ...

These coefficients are dependent on temperature. Some investigators also express the virial equation in terms of $1/V_m$:

$$\frac{p V_m}{RT} = 1 + \frac{B}{V_m} + \frac{C}{V_m^2} + \frac{D}{V_m^3} + \ldots \quad (7.30)$$

We have the relations:

$$B' = \frac{B}{RT} \quad C' = \frac{C - B^2}{R^2 T^2} \quad D' = \frac{2B^3 - 3BC + D}{R^3 T^3} \quad (7.31)$$

Fugacity of N_2 at 273 K f [bar]	0.999	96.8	405	2092
p [bar]	1	100	400	1000
	$\phi < 1$		$\phi > 1$	
	$\mu_{real} < \mu_{ideal}$		$\mu_{real} > \mu_{ideal}$	

Table 7.2   Fugacity of N_2 at 273 K and at various pressures.

Example

At 0°C for pressures up to 20 MPa, the virial coefficients for nitrogen gas are:

$B = -11.11 \cdot 10^{-6}$ m^3 mol^{-1} $B' = \dfrac{B}{RT} = -4.891 \cdot 10^{-9}$ Pa^{-1}

$C = 1798 \cdot 10^{-12}$ m^6 mol^{-2} $C' = \dfrac{C - B^2}{R^2 T^2} = 3.247 \cdot 10^{-16}$ Pa^{-2}

The compressibility factor is given by:

$Z = 1 - 4.891 \cdot 10^{-9} p + 3.247 \cdot 10^{-16} p^2$

The fugacity coefficient is obtained by integration. We find:

$\ln \phi = \displaystyle\int_0^p (Z-1)\dfrac{dp}{p}$

$= -4.891 \cdot 10^{-9} p + 1.623 \cdot 10^{-16} p^2$

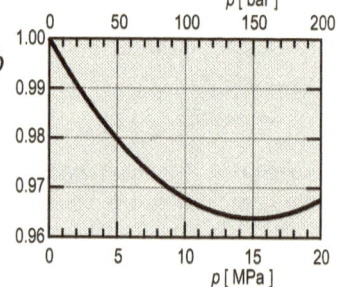

7.3.5   The van der Waals Equation of State

For n moles of gas, the van der Waals equation is:

$$p = \dfrac{nRT}{V-nb} - a\dfrac{n^2}{V^2} \quad \Leftrightarrow \quad \left(p + \dfrac{an^2}{V^2}\right)(V - nb) = nRT \quad (7.32)$$

The equation can also be written in terms of the molar volume, V_m:

$$\left(p + \dfrac{a}{V_m^2}\right)(V_m - b) = RT \quad (7.33)$$

As a real gas undergoes an isothermal compression, the observed behavior depends on the nature of the gas as well as on its temperature. We observe the existence of a temperature known as the **critical temperature**, which separates two domains where the behavior of the system is different. At a temperature lower than the critical temperature, we can observe, by compression, the condensation of the gas to a liquid phase. If we compress the gas at a temperature above the critical temperature, we never observe the appearance of a liquid phase. The critical temperature corresponds to the boundary between these two domains.

7 Thermodynamics of Gases

Fluid	a [Pa m^6 mol^{-2}]	b [m^3 mol^{-1}]	$\dfrac{R\,T_c}{p_c V_{c,m}}$
vdW gas	–	–	$\dfrac{8}{3} = 2.67$
He	3.460 10^{-3}	2.380 10^{-5}	3.34
Ne	2.080 10^{-2}	1.672 10^{-5}	3.18
H$_2$	2.453 10^{-2}	2.651 10^{-5}	3.26
O$_2$	1.382 10^{-1}	3.186 10^{-5}	3.49
N$_2$	1.370 10^{-1}	3.870 10^{-5}	3.44
CH$_4$	2.300 10^{-1}	4.301 10^{-5}	3.47

Table 7.3 Van der Waals equation constants a and b for a few gases (SI Units).

The curve corresponding to the critical temperature has an inflection point with a horizontal tangent. This point is referred to as the *critical point*. We obtain:

$$\left(\frac{\partial p}{\partial V_m}\right)_{T=T_c} = 0 \qquad \left(\frac{\partial^2 p}{\partial V_m^2}\right)_{T=T_c} = 0 \qquad (7.34)$$

$$\left.\begin{array}{l} p_c = \dfrac{a}{27\,b^2} \qquad V_{c,m} = 3\,b \qquad T_c = \dfrac{8\,a}{27\,R\,b} \\[6pt] \text{or } b = \dfrac{V_{c,m}}{3} \qquad a = 3\,p_c V_{c,m}^2 \qquad \text{with } \dfrac{R\,T_c}{p_c V_{c,m}} = \dfrac{8}{3} \end{array}\right\} \qquad (7.35)$$

Example

Using the van der Waals constants from table 7.3, we obtain the critical constants for nitrogen gas. We have:

$$p_c = \frac{a}{27\,b^2} = 3.39\ 10^6\ \text{Pa} \qquad T_c = \frac{8\,a}{27\,R\,b} = 126.1\ \text{K} \qquad V_{c,m} = 3b = 116\ \text{cm}^3$$

The critical pressure and temperature thus obtained are practically identical to the actual experimental values. The critical volume is actually 90 cm^3.

In table 7.3, we report the values of the quantity $R\,T_c/p_c V_{c,m}$. Often, as can be seen, real gases are not van der Waals gases. Let us write the van der Waals equation in terms of reduced variables:

$$\left.\begin{array}{l} p_r = \dfrac{p}{p_c} \qquad V_{r,m} = \dfrac{V_m}{V_{c,m}} \qquad T_r = \dfrac{T}{T_c} \\[6pt] \left(p_r + \dfrac{3}{V_{r,m}^2}\right)\left(V_{r,m} - \dfrac{1}{3}\right) = \dfrac{8}{3}\,T_r \end{array}\right\} \qquad (7.36)$$

Van der Waals gases, having the same values of their reduced variables are said to be in *corresponding states* and have identical properties. In Fig. 7.4, we display isothermal curves that for various values of the reduced temperature, T_r.

Below the critical temperature, the mathematical curves do not correspond to the experimental observations. From point A to point B, the van der Waals curve needs to be replaced by a horizontal line. The change going from the gaseous system A to the liquid system B isothermally takes place at constant pressure (isobaric).

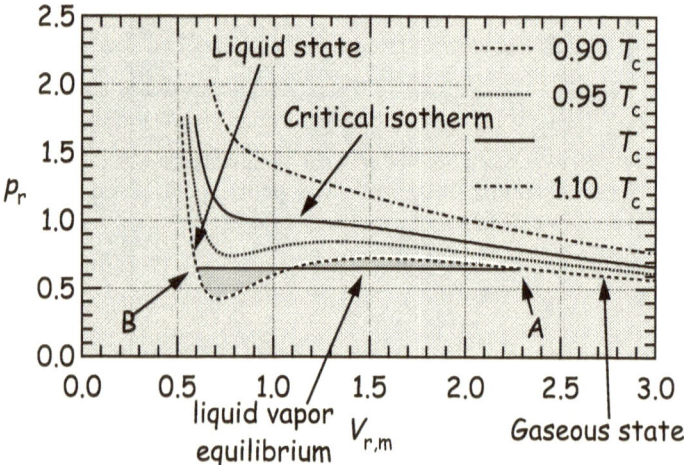

Figure 7.4 Van der Waals isotherms in reduced coordinates. The critical point has coordinates 1, 1.

The Gibbs energy of the system (equal to its chemical potential for one mole) is constant during the liquefaction process due to the two phase equilibrium. The expression for the thermodynamic differential of G, is the same as that for a system with a single phase (see chapter 8). We express the change in G between state A and state B.

$$\begin{aligned} dG &= V\,dp - S\,dT + \mu_\ell\,dn_\ell + \mu_g\,dn_g \\ &= V\,dp = d(p\,V) - p\,dV \\ &\text{since } \mu_\ell = \mu_g, \quad dn_\ell + dn_g = 0 \quad \text{and} \quad T = \text{constant} \\ G_B - G_A &= 0 = p(V_B - V_A) - \int_A^B p\,dV \quad \Rightarrow \quad p(V_{\ell,m} - V_{g,m}) = -w \end{aligned}$$

(7.37)

The area under the van der Waals curve between A and B is equal to the work done on the system during the process, $p(V_{g,m} - V_{\ell,m})$.

7 Thermodynamics of Gases

Example

For an ideal gas, the compressibility factor is always 1 in all conditions. We can calculate the compressibility factor for a van der Waals gas, using the reduced variables introduced above. We have:

$$Z = \frac{pV_m}{RT} = \frac{p_r V_{r,m} p_c V_{c,m}}{T_r \quad RT_c} = \frac{3}{8} \frac{p_r V_{r,m}}{T_r}$$

We represented graphically Z as a function of p_r, for two values of the reduced temperature. We see that, at the critical temperature, the compressibility factor is quite affected by pressure. For temperatures much above the critical temperature and low enough pressures, the compressibility factor is close to 1 (ideal gas behavior).

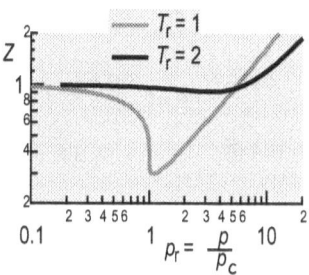

7.3.6 Joule-Thomson Effect

Consider a steady flow of gas at pressure p_I passing through a porous plug in a tube with *adiabatic* walls from which it comes out at pressure p_F. The experiment can be analyzed as a change from an initial state I to the final state F as shown in Fig. 7.5. We have:

$$q = 0 \Rightarrow U_F - U_I = w \tag{7.38}$$

The work done on the system is the sum of the work done by each piston.

$$w = -\int_{V_I}^{0} p_I dV - \int_{0}^{V_F} p_F dV = p_I V_I - p_F V_F \tag{7.39}$$

Using Eq. 7.38 and 7.39, we can conclude that:

$$U_F - U_I = p_I V_I - p_F V_F \Rightarrow U_F + p_F V_F = U_I + p_I V_I \Rightarrow H_F = H_I \tag{7.40}$$

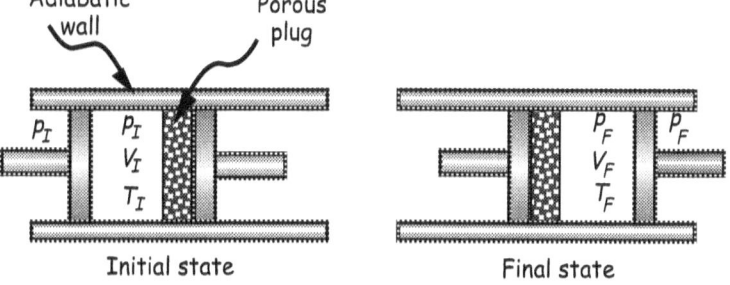

Figure 7.5 Joule-Thomson experiment. The walls of the tube and the pistons are adiabatic.

The enthalpy of the gas remains constant. Such a process is known as an *isenthalpic process*. The differential of the enthalpy is:

$$dH = \left(\frac{\partial H}{\partial p}\right)_T dp + \left(\frac{\partial H}{\partial T}\right)_p dT = V(1 - \alpha T)\, dp + C_p\, dT = 0 \quad (7.41)$$

The temperature change associated with a change in pressure at constant enthalpy is:

$$\mu_{JT} = \left(\frac{\partial T}{\partial p}\right)_H = \frac{V(\alpha T - 1)}{C_p} \quad (7.42)$$

μ_{JT} is the *Joule-Thomson coefficient*. For an ideal gas, α is $1/T$ and the Joule-Thomson coefficient is zero. A Joule-Thomson expansion of an ideal gas does not affect its temperature. In Fig. 7.6, we give a schematic representation of constant enthalpy curves for a real gas in a T vs. p diagram. Some of the curves, present a maximum. At these points, the Joule-Thomson coefficient, μ_{JT}, is zero. The corresponding temperature is the *inversion temperature*. The locus of these maxima, for all values of H, is the *inversion curve*.

A real gas can be cooled by a Joule-Thomson expansion (pressure decrease) if its representative point lies inside the inversion curve (Fig. 7.6). The expression of the inversion temperature for a van der Waals gas is obtained by equating the Joule-Thomson coefficient to zero. Using 5.43 and 7.41, we obtain:

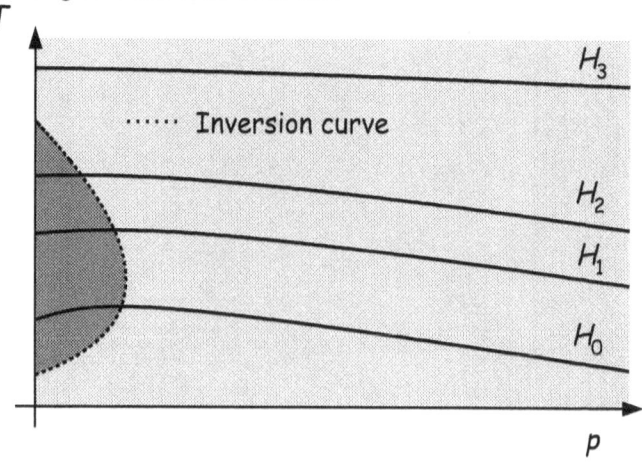

Figure 7.6 Lines of constant enthalpy of a gas and inversion temperature curve. Cooling by expansion can be achieved inside the inversion curve.

7 Thermodynamics of Gases

$$\left(\frac{\partial T}{\partial p}\right)_H = 0 \Rightarrow \left(\frac{\partial H}{\partial p}\right)_T = 0 = V - T\left(\frac{\partial V}{\partial T}\right)_p$$

$$V - T\left(\frac{\partial V}{\partial T}\right)_p = n \frac{\frac{2a}{V^2} - \frac{bRT}{(V-nb)^2}}{\frac{2an}{V^3} - \frac{RT}{(V-nb)^2}} \quad (7.43)$$

$$T_i = \frac{2a}{Rb}\left(\frac{V-nb}{V}\right)^2 = \frac{2a}{Rb}\left(\frac{V_m-b}{V_m}\right)^2$$

Example

We can use the value of T_i in the van der Waals equation to obtain the corresponding pressure in term of the molar volume. We have:

$$p = \frac{a}{b V_m^2}(2 V_m - 3b)$$

For Helium, $a = 3.46\ 10^{-3}$ Pa m^6 mol^{-2} and $b = 2.38\ 10^{-5}$ m^3 mol^{-1}. For a molar volume of $2\ 10^{-4}$ m^3, we find that the inversion temperature is 27.1 K and the corresponding pressure is $1.19\ 10^6$ Pa.

7.4 Mixtures of Real Gases

7.4.1 Chemical Potential of a Real Gas in a Mixture

In the expression of the chemical potential, the partial pressure of the gas is replaced by its fugacity:

$$\mu_i = \mu_i^\ominus(T) + RT \ln \frac{f_i}{p^\ominus} = \mu_i^\ominus(T) + RT \ln \frac{p_i}{p^\ominus} + RT \ln \phi_i$$

$$= \mu_i^\ominus(T) + RT \ln p + RT \ln x_i + RT \ln \phi_i \quad (7.44)$$

$$= \mu_{i\ \text{ideal}} + RT \ln \phi_i$$

with $\quad f_i = p_i\ \phi_i \quad$ and $\quad p_i = x_i\ p$

Species i has a fugacity f_i, a partial pressure p_i and its mole fraction in the mixture is x_i. The fugacity coefficient ϕ_i depends on T, p and the composition.

The total pressure of the system p is the sum of the partial pressures of the species. From these relations, we define the standard state of species i by:

$$\mu_i = \mu_i^\ominus(T) \text{ when } \begin{cases} p^\ominus = \text{standard state pressure (1 atm or 1 bar)} \\ x_i = 1 \text{ pure species} \\ \phi_i = 1 \text{ properties of an ideal gas} \end{cases}$$

(7.45)

7.4.2 Variables of Mixing for Real Gases

We express the partial molar properties starting from the expression for the chemical potential. Using 6.25 and 7.13 and the fact that $\mu_i^\ominus$ and x_i are independent of p:

$$\left. \begin{aligned} \overline{V}_i &= \left(\frac{\partial \mu_i}{\partial p}\right)_{T, n_i, n_j} = \left(\frac{\partial \mu_i^\ominus}{\partial p}\right)_{T, n_i, n_j} + \frac{RT}{p} \\ &\quad + RT\left(\frac{\partial \ln x_i}{\partial p}\right)_{T, n_i, n_j} + RT\left(\frac{\partial \ln \phi_i}{\partial p}\right)_{T, n_i, n_j} \\ \overline{V}_i &= V_{i, m}(\text{ideal}) + RT\left(\frac{\partial \ln \phi_i}{\partial p}\right)_{T, n_i, n_j} \end{aligned} \right\}$$

(7.46)

The volume of mixing, the global pressure remaining constant, is obtained as:

$$\left. \begin{aligned} \Delta_{mix} V &= \sum_i n_i \overline{V}_i - \sum_i n_i V_{i, m}(\text{real, pure}) \\ \text{with } V_{i, m}(\text{real, pure}) &= V_{i, m}(\text{ideal}) + RT\left(\frac{\partial \ln \phi_i(\text{pure})}{\partial p}\right)_T \end{aligned} \right\}$$

(7.47)

Note $\phi_i(\text{pure}) \neq \phi_i$ in the mixture. The volume of mixing is:

$$\Delta_{mix} V = \sum_i n_i RT \left[\left(\frac{\partial \ln \phi_i}{\partial p}\right)_{T, n_i, n_j} - \left(\frac{\partial \ln \phi_i(\text{pure})}{\partial p}\right)_T \right]$$

(7.48)

The enthalpy of mixing is obtained in a very similar way, using Eq. 6.29:

$$\Delta_{mix} H = \sum_i n_i \overline{H}_i - \sum_i n_i H_{i, m}(\text{real, pure})$$

(7.49)

7 Thermodynamics of Gases

$$\overline{H}_i = -T^2 \frac{\partial}{\partial T}\left(\frac{\mu_i}{T}\right)_{p, n_i, n_j} = H_{i, m} \text{ (ideal)} - RT^2 \left(\frac{\partial \ln \phi_i}{\partial T}\right)_{p, n_i, n_j} \tag{7.50}$$

$$\Delta_{mix}H = \sum_i n_i R T^2 \left\{ \left(\frac{\partial \ln \phi_i(\text{pure})}{\partial T}\right)_p - \left(\frac{\partial \ln \phi_i}{\partial T}\right)_{p, n_i, n_j} \right\} \tag{7.51}$$

The internal energy of mixing is:

$$\Delta_{mix}U = \Delta_{mix}H - p\Delta_{mix}V \tag{7.52}$$

The difference between the mixing variable and that of the corresponding ideal gas mixture, is called the **excess variable of mixing**. The excess Gibbs energy of mixing is:

$$G^E = \Delta_{mix}G(\text{real}) - \Delta_{mix}G(\text{ideal}) = \sum_i n_i R T \ln\left\{\frac{\phi_i}{\phi_i(\text{pure})}\right\} \tag{7.53}$$

Other excess variables can be obtained for other extensive variables, such as the **excess entropy of mixing**. We first express the entropy of mixing of the real gases. It is given by:

$$\Delta_{mix}S(\text{real}) = \sum_i n_i \overline{S}_i - \sum_i n_i S_{i, m}(\text{real, pure}) \tag{7.54}$$

We need to evaluate each of the terms. We have:

$$\left.\begin{array}{l} \overline{S}_i = -\left(\dfrac{\partial \mu_i}{\partial T}\right)_{p, n_i, n_j} \\[4pt] = S_i^\ominus - R\ln p - R\ln x_i - R\ln \phi_i - RT\left(\dfrac{\partial \ln \phi_i}{\partial T}\right) \\[4pt] = S_{i, m}(\text{ideal}) - R\ln \phi_i - RT\left(\dfrac{\partial \ln \phi_i}{\partial T}\right)_{p, n_i, n_j} \end{array}\right\} \tag{7.55}$$

$$S_{i, m}(\text{real, pure}) = S_{i, m}(\text{ideal}) - R\ln \phi_i(\text{pure}) - RT\left(\frac{\partial \ln \phi_i}{\partial T}\right)_{p, n_i, n_j} \tag{7.56}$$

$$\Delta_{mix} S(real) = -\sum_i n_i R \left[\ln x_i + \ln \left\{ \frac{\phi_i}{\phi_i(pure)} \right\} \right.$$
$$\left. + T \left(\frac{\partial \ln \left\{ \frac{\phi_i}{\phi_i(pure)} \right\}}{\partial T} \right)_{p, n_i, n_j} \right] \quad (7.57)$$

The excess entropy of mixing is:

$$S^E = \Delta_{mix} S(real) - \Delta_{mix} S(ideal)$$

$$= -\sum_i n_i R \left[\ln \left\{ \frac{\phi_i}{\phi_i(pure)} \right\} + T \left(\frac{\partial \ln \left\{ \frac{\phi_i}{\phi_i(pure)} \right\}}{\partial T} \right)_{p, n_i, n_j} \right] \quad (7.58)$$

Note:

$$S^E = -\left(\frac{\partial G^E}{\partial T} \right)_{p, n_i, n_j} \quad (7.59)$$

7.5 Ideal Mixtures of Gases

A **gas mixture** (of real gases) is an **ideal mixture** if the chemical potential of the species of the mixture can be expressed as:

$$\mu_i = \mu_i^*(T, p) + RT \ln x_i \quad (7.60)$$

It differs from an *ideal gas mixture* by the fact that the dependence of the chemical potential on p is not defined.

Example

Consider an ideal mixture of gases. For such a mixture, many variables of mixing have very simple expressions. The partial volumes of the species in the mixture are independent of the composition and equal to their molar volumes.

$$\overline{V}_i = \left(\frac{\partial \mu_i}{\partial p} \right)_{T, n_i, n_j} = \left(\frac{\partial \mu_i^*}{\partial p} \right)_T = V_{i, m} \Rightarrow V = \sum_i n_i \overline{V}_i = \sum_i n_i V_{i, m}$$

The partial molar enthalpies are simply the molar quantities of the pure species.

$$\left. \begin{array}{l} \overline{H}_i = -T^2 \left[\frac{\partial}{\partial T} \left(\frac{\mu_i}{T} \right) \right]_{p, n_i, n_j} = -T^2 \left[\frac{\partial}{\partial T} \left(\frac{\mu_i^*}{T} \right) \right]_p = H_{i, m} \\ \Downarrow \\ H = \sum_i n_i \overline{H}_i = \sum_i n_i H_{i, m} \end{array} \right\}$$

7 Thermodynamics of Gases

The partial molar internal energies are also equal to the molar internal energies of the pure species.

$$\left.\begin{array}{c} U_{i,m} = H_{i,m} - p V_{i,m} \quad \text{and} \quad \overline{U}_i = \overline{H}_i - p \overline{V}_i \\ \Downarrow \\ U = H - pV = \sum_i n_i H_{i,m} - p \sum_i n_i V_{i,m} = \sum_i n_i (H_{i,m} - p V_{i,m}) \\ = \sum_i n_i U_{i,m} = \sum_i n_i \overline{U}_i \end{array}\right\}$$

Here the volume of mixing, $\Delta_{mix}V$, the enthalpy of mixing, $\Delta_{mix}H$, and the internal energy of mixing $\Delta_{mix}U$, are zero for an ideal mixture of gases.

For a system at pressure p and temperature T, the chemical potential of a gas in an *ideal mixture* can be expressed using either 7.44 or 7.60, which need to have the same value:

$$\left.\begin{array}{c} \mu_i = \mu_i^{\ominus}(T) + R T \ln f_i = \mu_i^*(T,p) + R T \ln x_i \\ \Downarrow \\ R T \ln \dfrac{f_i}{x_i} = \mu_i^* - \mu_i^{\ominus} \end{array}\right\} \quad (7.61)$$

The difference $\mu_i^* - \mu_i^{\ominus}$ depends only on p and T and does not depend on the composition of the system. As a consequence, the ratio f_i/x_i is constant for any value of x_i. We can write:

$$f_i = f_i^* x_i \qquad (7.62)$$

where f_i^* is the fugacity of pure gas i at temperature T and pressure p. Eq. 7.62 is known as the **Lewis-Randall rule**. For systems where this rule is valid, it is only necessary to have data about the pure component to obtain the fugacities of in an ideal mixture.

Example

For pure nitrogen and methane, we have the following data at 20 MPa and 298 K.

$f_{N_2} = 1.86 \cdot 10^7$ Pa $\quad f_{CH_4} = 1.42 \cdot 10^7$ Pa

The fugacities at 20 MPa and 298 K in a mixture where $x_{CH_4} = 0.7$ can be estimated to be:

$f_{N_2} \simeq 0.3 \cdot 1.86 \cdot 10^7 \simeq 0.56 \cdot 10^7$ Pa $\quad f_{CH_4} \simeq 0.7 \cdot 1.42 \cdot 10^7 \simeq 0.99 \cdot 10^7$ Pa

8. Systems with Several Phases, No Chemical Reaction. Third Law of Thermodynamics

8.1 Introduction

A *phase* is a part of a system which, *on a macroscopic scale*, can be considered as *homogeneous*. A closed system, made of several phases, with species that can freely move from one phase to another, can be considered as an ensemble of open subsystems, with only one phase each.

8.2 Differential Expressions for State Functions

Let us envisage a *closed system* where *no chemical reaction* can take place and several phases are present at equilibrium. Each of the phases is, by itself, a subsystem in a state of equilibrium. The differentials of the internal energy and of the Gibbs energy for phase α are:

$$\left. \begin{array}{l} dU^\alpha = -p^\alpha \, dV^\alpha + T^\alpha \, dS^\alpha + \sum_i \mu_i^\alpha \, dn_i^\alpha \\ \\ dG^\alpha = V^\alpha \, dp^\alpha - S^\alpha \, dT^\alpha + \sum_i \mu_i^\alpha \, dn_i^\alpha \end{array} \right\} \quad (8.1)$$

Mechanical and thermal equilibria imply:

$$p^\alpha = p^\beta = p^\gamma = p \quad \text{and} \quad T^\alpha = T^\beta = T^\gamma = T \quad (8.2)$$

Example

Consider a closed system containing a single species in contact with a thermal reservoir, with its internal pressure uniform, constant and equal to the external pressure acting on it. We assume that a liquid phase and a gaseous phase are present. We can find the criterion for the liquid and the gas phase to be at equilibrium. Since the pressure and temperature of both phases are the same in view of the mechanical and thermal equilibrium, the expression for the differential of the Gibbs energy is:

$$dG = V \, dp - S \, dT + \mu_\ell \, dn_\ell + \mu_g \, dn_g$$

The temperature and pressure are constant and the number of moles of the species is constant as well.
The condition for the system to be at equilibrium is that G has reached its minimum value. We have :

$$\left.\begin{array}{l} n_\ell + n_g = 1 \Rightarrow dn_g = -dn_\ell \\ \Downarrow \\ dG = (\mu_\ell - \mu_g)\, dn_\ell \end{array}\right\} \text{Equilibrium} \Rightarrow dG = 0 \Rightarrow \mu_\ell = \mu_g$$

At equilibrium, the chemical potential of the species is the same in both phases.

It can be shown that, in general, equilibrium implies that the chemical potential of a species is the same in all phases.

$$\mu_i^\alpha = \mu_i^\beta = \mu_i^\gamma = \ldots = \mu_i \tag{8.3}$$

The differentials of the internal energy and of the Gibbs energy of a system, where phase equilibrium is achieved, are :

$$\left.\begin{array}{l} dU = -p\, dV + T\, dS + \sum_i \mu_i\, dn_i \\ \\ dG = V\, dp - S\, dT + \sum_i \mu_i\, dn_i \end{array}\right\} \tag{8.4}$$

The expressions are identical to those obtained for systems with one single phase. We have the relations :

$$\left.\begin{array}{l} dV = \sum_\alpha dV^\alpha \quad dS = \sum_\alpha dS^\alpha \quad V = \sum_\alpha V^\alpha \\ \\ S = \sum_\alpha S^\alpha \quad dn_i = \sum_i dn_i^\alpha \end{array}\right\} \tag{8.5}$$

8.3 Spontaneous Transfer of a Species from One Phase to Another One

Consider a system that can only exchange volume work with its surroundings and contains two phases α and β. At *constant pressure and temperature*, dn_i moles of species i spontaneously go from phase α to phase β. For this process to be spontaneous, the corresponding change of the Gibbs energy needs to be negative (§ 5.7).

$$dn_i^\beta = -dn_i^\alpha > 0 \quad \text{and} \quad dG = (-\mu_i^\alpha + \mu_i^\beta)\,dn_i^\beta < 0$$
$$\Downarrow$$
$$\mu_i^\alpha > \mu_i^\beta \tag{8.6}$$

A chemical species tends to spontaneously move to another phase in which its chemical potential is smaller.

Example

Consider a system in contact with a thermal reservoir at 0°C, where liquid water and ice are at equilibrium under a pressure of 1 bar. The molar volume of liquid water and ice are respectively $V_{m(l)} = 18.02\ 10^{-6}$ m³ mol⁻¹ and $V_{m(s)} = 19.66\ 10^{-6}$ m³ mol⁻¹. We increase the pressure to 10 bar. The change in the chemical potential of water and ice due to the pressure increase are obtained using Eq. 5.31 (we neglect the molar volume changes of water or ice due to the pressure change). We have:

$$\left(\frac{\partial \mu}{\partial p}\right)_{T,n} = \left(\frac{\partial V}{\partial n}\right)_{T,p} = V_m$$
$$\Downarrow$$
$$\mu_{water}(10\text{ bar}) - \mu_{water}(1\text{ bar}) = 18.02\ 10^{-6} \cdot (10^6 - 10^5) = 16.22\text{ J mol}^{-1}$$
$$\mu_{ice}(10\text{ bar}) - \mu_{ice}(1\text{ bar}) = 19.66\ 10^{-6} \cdot (10^6 - 10^5) = 17.69\text{ J mol}^{-1}$$

From this result we find that at 0°C and 10 bar, the chemical potential of ice is larger that the chemical potential of water.

$$\mu_{ice}(10\text{ bar}) - \mu_{water}(10\text{ bar}) = 1.47\text{ J mol}^{-1}$$

We can conclude, using the results of Eq. 8.6, that ice will transform into water. Liquid water is the stable phase at 0°C and 10 bar.

8.4 The Phase Rule

Consider a *closed system, at equilibrium*, with *several phases and several chemical species*. The number of *intensive variables*, which can be arbitrarily modified, is the **number of degrees of freedom**, or **variance**, v, of the system. For *each phase*, the intensive variables are:

$$T, p, \ldots, x_i, \ldots \tag{8.7}$$

its temperature, its pressure and the mole fraction of its species. These variables are not independent for each of the phases of the system. Let n be the **number of chemical species** of the system present in every phase. There are $n + 2$ intensive variables that are needed to characterize each phase. Let φ be the **number of phases** in the system. We need to know these variables for each of the phases, so that altogether we must know $\varphi(n+2)$ variables. A number of relations exist that we must take into account:

- The sum of the mole fractions of all of the species present in each phase is 1. $\Rightarrow \varphi$ equations.

- Thermal and mechanical equilibrium. The temperature is the same in all phases and pressure is uniform in all phases. $\Rightarrow 2(\varphi-1)$ equations.

- The chemical potential of each species is the same in each phase. $\Rightarrow n(\varphi-1)$ equations.

We obtain for the variance v:

$$\left. \begin{array}{c} v = \varphi(n+2) - \varphi - 2(\varphi-1) - n(\varphi-1) \\ \Downarrow \\ v = n + 2 - \varphi \end{array} \right\} \quad (8.8)$$

This relation constitutes the **phase rule**. One can show that it remains valid even if one of the species is not present in one of the phases.

Example

The variance of a system where liquid water is in equilibrium with water vapor is:
$v = 1 + 2 - 2 = 1$
from this relation one can conclude that for a given value of temperature, the pressure of the vapor in a system at equilibrium is determined.
The variance of a system where CO_2 solid is in equilibrium with CO_2 vapor is:
$v = 1 + 2 - 2 = 1$
From this relation one can conclude that for a given value of pressure, the temperature of the system at equilibrium is determined.

8.5 Equilibrium of Two Phases of a Pure Substance

8.5.1 Clapeyron Equations

Using the phase rule, we find that a system, where two phases α and β of a pure substance are at equilibrium, is monovariant:

$$v = n + 2 - \varphi = 1 + 2 - 2 = 1 \qquad (8.9)$$

If the temperature of such a system is known, its equilibrium pressure is determined and reciprocally. The curve on Fig. 8.1 represents the equilibrium pressure of two phases α and β as a function of temperature. At temperature T_1, the two phases are at equilibrium only when the system pressure is p_1.

8 Systems with Several Phases, No Chemical Reaction. Third Law...

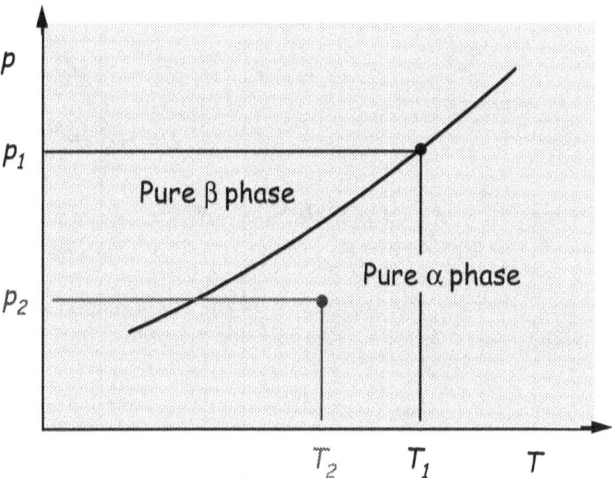

Figure 8.1 Equilibrium curve of two phases α and β.

For an infinitesimal change from an equilibrium state at T, p to a new equilibrium state at $T + dT, p + dp$, we have:

$$\left.\begin{array}{l}\mu^\alpha(T, p) = \mu^\beta(T, p) \\ \mu^\alpha(T + dT, p + dp) = \mu^\beta(T + dT, p + dp)\end{array}\right\} \quad (8.10)$$

$$\left(\frac{\partial \mu^\alpha}{\partial T}\right)_p dT + \left(\frac{\partial \mu^\alpha}{\partial p}\right)_T dp = \left(\frac{\partial \mu^\beta}{\partial T}\right)_p dT + \left(\frac{\partial \mu^\beta}{\partial p}\right)_T dp \quad (8.11)$$

Replace the partial derivatives by their expressions in terms of molar quantities using Eq. 6.30:

$$\left.\begin{array}{c}-S_m^\alpha \, dT + V_m^\alpha \, dp = -S_m^\beta \, dT + V_m^\beta \, dp \\ \Downarrow \\ \dfrac{dp}{dT} = \dfrac{S_m^\alpha - S_m^\beta}{V_m^\alpha - V_m^\beta}\end{array}\right\} \quad (8.12)$$

We can also write:

$$\left.\begin{array}{c}\mu^\alpha = H_m^\alpha - TS_m^\alpha = \mu^\beta = H_m^\beta - TS_m^\beta \\ \Downarrow \\ S_m^\alpha - S_m^\beta = \dfrac{H_m^\alpha - H_m^\beta}{T}\end{array}\right\} \Rightarrow \left\{\begin{array}{c}\dfrac{dp}{dT} = \dfrac{H_m^\alpha - H_m^\beta}{T\left(V_m^\alpha - V_m^\beta\right)} \\ L = H_m^\alpha - H_m^\beta\end{array}\right.$$

(8.13)

The difference $H_m^\alpha - H_m^\beta$ is the **enthalpy of the phase change**, or **latent heat** of the phase change, symbol L. Latent heats are always taken as positive. Equations 8.12 and 8.13 are the **Clapeyron equations**. At T_2 and p_2, the variance v of the system must be 2. The number of phases can only be 1. The only phase present is phase α.

Example

At 200°C the latent heat of vaporization of water is 34945 J mol^{-1}. The molar volume of water vapor at that temperature is 2.2913 10^{-3} m^3 mol^{-1}. The molar volume of liquid water is 20.835 10^{-6} m^3 mol^{-1}. The slope of the equilibrium curve at this temperature is:

$$\frac{dp}{dT} = \frac{L}{T(V_m(\text{vap}) - V_m(l))} = \frac{34945}{473.15 \cdot (2.2913\ 10^{-3} - 20.835\ 10^{-6})} = 32529\ \text{Pa K}^{-1}$$

The molar entropy change from liquid to vapor is:

$$S_m(\text{vap}) - S_m(l) = \frac{L}{T} = \frac{34945}{473.15} \simeq 73.85\ \text{J mol}^{-1}\ \text{K}^{-1}.$$

8.5.2 Equilibrium between a Gaseous Phase and a Condensed Phase (Liquid or Solid) of a Pure Substance

If the phase α is an ideal gas, then the molar volume of the solid phase can often be neglected, leading to the **Clausius-Clapeyron equation**:

$$V_m^\alpha - V_m^\beta \simeq V_m(g) \simeq \frac{RT}{p} \Rightarrow \frac{d\ln p}{dT} \simeq \frac{H_m(g) - H_m(\text{cond})}{RT^2} = \frac{L}{RT^2}$$

(8.14)

Example

At 170 K, solid CO_2 can be in equilibrium with gaseous CO_2

$$CO_2\ (\text{solid}) \rightleftharpoons CO_2\ (g)$$

The slope of the sublimation curve of CO_2 on a ln p vs. T graph (using SI units) is:

$$\frac{d\ln p}{dT} \simeq 0.109\ \text{K}^{-1}$$

The latent heat of sublimation of CO_2 is:

$$L = \frac{d\ln p}{dT} \cdot R \cdot T^2 = 0.109 \cdot 8.3145 \cdot 170^2 \simeq 26200\ \text{J mol}^{-1}$$

8 Systems with Several Phases, No Chemical Reaction. Third Law...

8.5.3 Schematic Representation of some of the Thermodynamic Functions in the Vicinity of a Phase Change

When two phases, α and β, of a pure substance are at equilibrium, the chemical potential of the substance is the same in both phases. The variations of some thermodynamic functions, at constant temperature or constant pressure, are expressed by:

$$\left.\begin{array}{l}\left(\dfrac{\partial \mu}{\partial p}\right)_T = V_m \qquad \left(\dfrac{\partial \mu}{\partial T}\right)_p = -S_m \\[2ex] \left(\dfrac{\partial^2 \mu}{\partial p^2}\right)_T = \left(\dfrac{\partial V_m}{\partial p}\right)_T = -\kappa V \qquad \left(\dfrac{\partial^2 \mu}{\partial T^2}\right)_p = -\left(\dfrac{\partial S_m}{\partial T}\right)_p = -\dfrac{C_{p,m}}{T}\end{array}\right\} \quad (8.15)$$

In Fig. 8.2, we have a schematic representation of the changes in these functions with either pressure or temperature under isothermal or isobaric conditions respectively. All of them present discontinuities of their slopes or values.

8.5.4 Effect of the Pressure of an Insoluble Gas on the Vapor Pressure of a Liquid

We consider a system where pressure is exerted on the liquid by a gas assumed to be insoluble[†]. Vapor and liquid are at equilibrium.

$$\mu(l) = \mu(vapor) = \mu^{\ominus}(vapor)\,(T) + R\,T\ln p\,(vapor) \qquad (8.16)$$

where $p\,(vapor)$ is the partial pressure of the liquid phase in the presence of the gas, the total pressure is p. The standard chemical potential $\mu^{\ominus}(vapor)$ depends only on temperature. Using Eq. 6.30:

$$\left.\begin{array}{l}\left(\dfrac{\partial \mu(l)}{\partial p}\right)_T = \overline{V}(l) = V_m(l) = R\,T\left(\dfrac{\partial \ln p(vapor)}{\partial p}\right)_T \\[2ex] \Downarrow \\[1ex] \left(\dfrac{\partial \ln p(vapor)}{\partial p}\right)_T = \dfrac{V_m(l)}{R\,T}\end{array}\right\} \quad (8.17)$$

The molar volume of a liquid is often small compared to $R\,T$ and the partial vapor pressure is only slightly modified by the presence of an inert gas.

[†] If the gas is soluble in the liquid, then the thermodynamics of solutions must be used (See chapter 12).

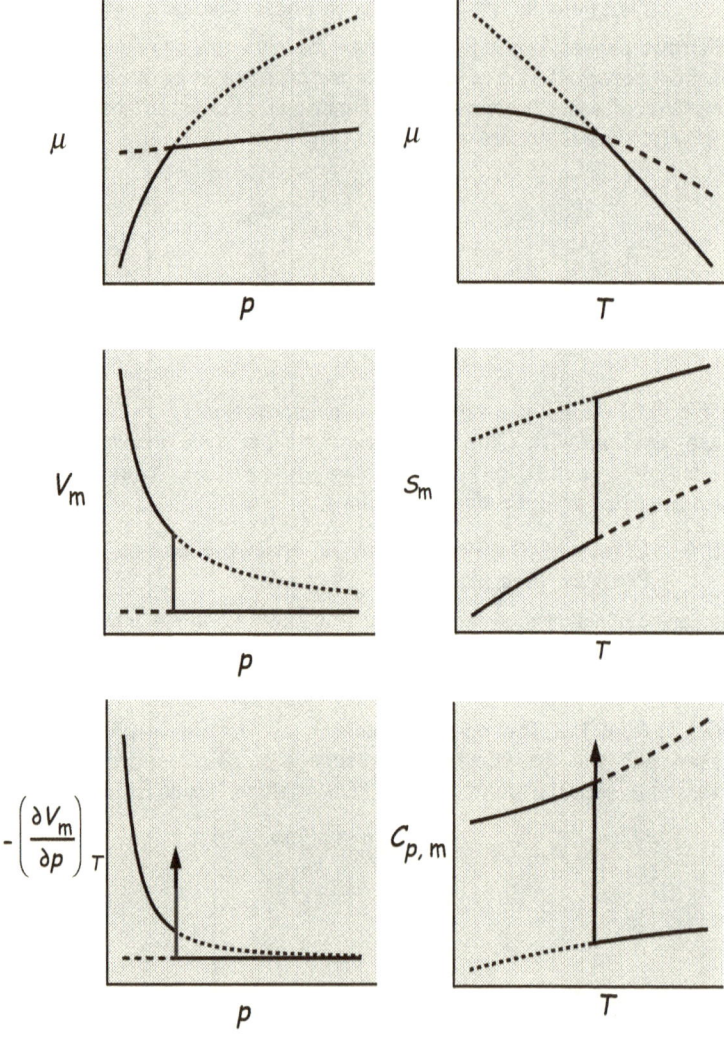

Figure 8.2 Schematic representation of various thermodynamic functions in the vicinity of a phase change of a pure substance. The continuous lines correspond to stable phases, dashed lines correspond to metastable phases (when they exist).

8 Systems with Several Phases, No Chemical Reaction. Third Law...

Example
At 313 K, the molar volume of water is 18.14 cm³ and its vapor pressure is 7.38 kPa. By integration of Eq. 8.17, we obtain:

$$\ln \frac{p(\text{vapor with gas})}{p(\text{vapor pure})} = \frac{V_m^{(l)}}{RT} \left(p(\text{vapor with gas}) + p(g) - p(\text{vapor pure}) \right)$$

Let us find the effect of a 10^7 Pa air pressure. We calculate the right hand side approximately. We have:

$$\left. \begin{array}{l} p(\text{vapor with gas}) + p(g) - p(\text{vapor pure}) \simeq p(g) \\ \ln \frac{p(\text{vapor with gas})}{7.38 \cdot 10^3} \simeq \frac{18.14 \cdot 10^{-6}}{8.3145 \cdot 313} p(g) = 0.0697 \end{array} \right\} \Rightarrow p(\text{vapor with gas}) \simeq 7913 \text{ Pa}$$

The vapor pressure is increased by 7%. This result is practically identical to a numerical solution of the complete equation.

8.5.5 Effect of Temperature on the Latent Heat of Phase Change and on the Equilibrium Pressure

We write the differential of the latent heat (as a function of T and p) from its definition in Eq. 8.13 and make use of relations 5.7 and 5.44. This leads to:

$$\left. \begin{array}{l} dL = \left[\frac{\partial \left(H_m^\alpha - H_m^\beta \right)}{\partial T} \right]_p dT + \left[\frac{\partial \left(H_m^\alpha - H_m^\beta \right)}{\partial p} \right]_T dp \\ = \left(C_{p,m}^\alpha - C_{p,m}^\beta \right) dT + \left[V_m^\alpha - T \left(\frac{\partial V_m^\alpha}{\partial T} \right)_p - V_m^\beta + T \left(\frac{\partial V_m^\beta}{\partial T} \right)_p \right] dp \\ = \Delta C_{p,m} dT + \left[\Delta V_m - T \left(\frac{\partial \Delta V_m}{\partial T} \right)_p \right] dp \end{array} \right\} \quad (8.18)$$

We replace dp by its value in terms of dT using Eq. 8.13:

$$\left. \begin{array}{l} \frac{dL}{dT} = \Delta C_{p,m} + \left[\Delta V_m - T \left(\frac{\partial \Delta V_m}{\partial T} \right)_p \right] \frac{L}{T \Delta V_m} \\ = \Delta C_{p,m} + \frac{L}{T} - L \left(\frac{\partial \ln \Delta V_m}{\partial T} \right)_p \end{array} \right\} \quad (8.19)$$

If phase α is an ideal gas and phase β is a liquid or a solid, we have the approximate relation:

$$\Delta V_m = V_m^\alpha - V_m^\beta \simeq V_m(\text{gas}) = \frac{RT}{p} \Rightarrow L \left(\frac{\partial \ln \Delta V_m}{\partial T} \right)_p \simeq \frac{L}{T} \quad (8.20)$$

which leads to a simple approximate result:

$$\frac{dL}{dT} \simeq \Delta C_{p,m} \quad (8.21)$$

Example

At 298.15 K, from the molar heat capacities for liquid water and water vapor, we obtain :

$$\left. \begin{array}{l} C_{p,m}(\text{vap}) = 33.56 \text{ J mol}^{-1} \text{ K}^{-1} \\ C_{p,m}(l) = 75.29 \text{ J mol}^{-1} \text{ K}^{-1} \end{array} \right\} \Rightarrow \frac{dL}{dT} = -41.7 \text{ J mol}^{-1} \text{ K}^{-1}$$

From the latent heat of vaporization at 273 K and 313 K, we find :

$$\left. \begin{array}{l} L_{273} = 45054 \text{ J mol}^{-1} \\ L_{313} = 43350 \text{ J mol}^{-1} \end{array} \right\} \Rightarrow \frac{dL}{dT} = \frac{43350 - 45054}{313 - 273} = -42.6 \text{ J mol}^{-1}\text{K}^{-1}$$

These two results are in fairly good agreement.

8.6 Phase Diagram of a Pure Substance

In Fig. 8.3 we display a phase diagram for a pure substance can exist as a gas, a liquid or a solid under one crystalline form. The equilibrium of two phases corresponds to one line. Delimited by the equilibrium lines, are domains that correspond to stability regions for single phases.

We have three different curves representing the possible equilibria of the different phases that can be observed. The vaporization curve corresponds to the liquid-vapor equilibrium. The fusion curve corresponds to the solid-liquid equilibrium. Finally, the sublimation curve corresponds to the solid-vapor equilibrium. The fusion curve and the vaporization curve cross in some point, where we can write :

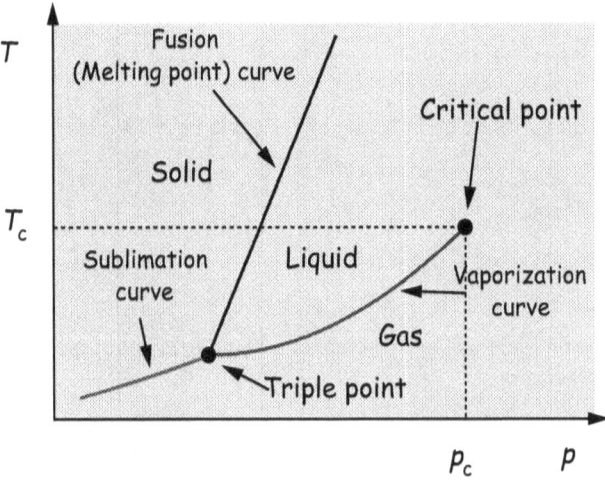

Figure 8.3 Phase diagram of a pure substance. Continuity of the fluid state. Triple point. Critical point.

8 Systems with Several Phases, No Chemical Reaction. Third Law...

$$\left.\begin{array}{l}\mu(\text{liquid}) = \mu(\text{vapor})\\ \mu(\text{solid}) = \mu(\text{liquid})\end{array}\right\} \Rightarrow \mu(\text{solid}) = \mu(\text{vapor}) \qquad (8.22)$$

In that point, the chemical potential of the solid is the same as that of the vapor, thus that point must also be on the sublimation curve. The three curves meet in a single point known as the **triple point** of the substance. At that particular point, the variance of the system is:

$$v = 1 + 2 - 3 = 0 \qquad (8.23)$$

A system with variance 0 is said to be **invariant**. As an example, we have already mentioned that the triple point of water is characterized by $T_{tp\ H_2O} = 273.16$ K and $p_{tp\ H_2O} = 6.11$ mbar.

Example
At the triple point, the latent heat of fusion, vaporization and sublimation are related. We have:

$$\left.\begin{array}{l}L_{vap} = H_{m\ vap} - H_{m\ liq}\\ L_{fus} = H_{m\ liq} - H_{m\ sol}\end{array}\right\} \Rightarrow L_{sub} = H_{m\ vap} - H_{m\ sol} = L_{vap} + L_{fus}$$

From the values for water $L_{vap} = 45.0$ kJ mol^{-1} and $L_{fus} = 6.0$ kJ mol^{-1}, we find that the latent heat of sublimation of ice is $L_{sub} = 51$ kJ mol^{-1}.

The vaporization curve stops at the **critical point**. At temperatures above the critical temperature, T_c, liquefaction does not take place by isothermal compression of the system. When the system pressure is above the critical pressure p_c, during an isobaric cooling, liquefaction does not take place either. If both, the temperature and pressure, are above the critical values, only one fluid state is observed.
The properties of the gas phase and of the liquid phase get close to one another in the vicinity of the critical point. The slope of the vaporization curve remains finite at the critical point. Equation 8.13 implies that the numerator and the denominator both tend to zero simultaneously. Near the critical point, we have the following relations:

$$\left.\begin{array}{l}H_m(\text{liquid}) \simeq H_m(\text{vapor})\\ S_m(\text{liquid}) \simeq S_m(\text{vapor})\\ V_m(\text{liquid}) \simeq V_m(\text{vapor})\end{array}\right\} \qquad (8.24)$$

Under certain conditions, it is possible to observe some delay to an expected phase change.

8.7 Evaluation of Entropies

Using our present knowledge, we can evaluate the change of the entropy of a species as a function of temperature assuming that we know its entropy at some temperature T_0. During an isobaric process, we can evaluate the entropy change using Eq. 5.42 :

$$S(T) = S(T_0) + \int_{T_0}^{T} \frac{C_p}{T} \, dT \tag{8.25}$$

If the system undergoes some phase transition (isobaric condition), as the system receives heat, its temperature stays constant while the phase change takes place. The entropy change during a fusion, for example, is obtained from the enthalpy of fusion, ΔH_{fus} at T_{fus}. When a fusion takes place, at constant pressure and T_{fus}, we find :

$$S(T) = S(T_0) + \int_{T_0}^{T_{fus}} \frac{C_p(s)}{T} \, dT + \frac{\Delta H_{fus}}{T_{fus}} + \int_{T_{fus}}^{T} \frac{C_p(l)}{T} \, dT \tag{8.26}$$

provided heat capacities and latent heats of phase change are known. Additional terms to Eq. 8.26 are necessary if vaporization takes place.

Example

We can obtain the entropy change corresponding to the vaporization of 0.1 mole of water at 25°C in equilibrium with its vapor at 3167 Pa. The latent heat of vaporization is $L_{vap} = 43990$ J mol^{-1}. The entropy change is :

$$S(vap) - S(l) = \frac{q_{rev}}{T} = \frac{nL_{vap}}{T} = \frac{0.1 \cdot 43990}{298.15} \simeq 14.75 \text{ J}$$

For a process where V,T or p,T are variable, we have (taking into account Maxwell's relations 5.28 and 5.29 and relations 5.36 or 5.42) :

$$dS = \frac{C_V}{T} \, dT + \left(\frac{\partial p}{\partial T}\right)_V dV = \frac{C_p}{T} \, dT - \left(\frac{\partial V}{\partial T}\right)_p dp \tag{8.27}$$

The entropy change between two states is obtained by integrating whichever is the most convenient of these expressions. It is necessary to select a value $S(T_0)$ because the changes of the Helmholtz energy and of the Gibbs energy depends on the value selected for $S(T_0)$.

The selected reference state for entropy must have *definite physical significance*, with respect to all other possible choices. The choice must be of universal character for any existing substance.

8.8 Third Law of Thermodynamics

In view of the experimental behavior of the thermodynamic state functions A and G near 0 K, this temperature has been chosen as the reference temperature for entropy. Historically, Nernst stated that the entropy of any pure substance is finite at 0 K. Subsequently, Planck completed this law by stating that the finite value is zero for all pure species in a perfect crystalline state. The third law of thermodynamics can be stated the following way :

> The entropy of any pure species in a perfect crystalline state is finite at 0 K. By convention, this finite value is taken to be zero for all chemical species.

8.9 Implications of the Third Law

8.9.1 Heat Capacities

Selecting 0 K as the reference temperature is only possible if all of the heat capacities tend to zero when temperature approaches 0 K, in such a way that expressions 8.25 and 8.26 may be evaluated. Experimentally, one observes that heat capacities do tend towards zero as $T \to 0$ K. Quantum mechanical considerations show that heat capacities of solids vary according to a T^3 law near 0 K.

8.9.2 Effect of Pressure and Volume on Entropy at 0 K

The isobaric coefficient of thermal expansion α becomes very small as temperature approaches 0 K. Using 5.29, we find :

$$\left(\frac{\partial S}{\partial p}\right)_T = -\left(\frac{\partial V}{\partial T}\right)_p = 0 \Rightarrow \left(\frac{\partial S}{\partial p}\right)_{T=0} = 0 \qquad (8.28)$$

which shows that entropy near 0 K is independent of pressure. Experimentally, it is found that the isothermal compressibility coefficient κ (Eq. 1.6) remains finite in the vicinity of 0 K. Using Eq. 5.53, we obtain :

$$\left(\frac{\partial S}{\partial V}\right)_T = \frac{\left(\frac{\partial S}{\partial p}\right)_T}{\left(\frac{\partial V}{\partial p}\right)_T} = 0 \Rightarrow \left(\frac{\partial S}{\partial V}\right)_{T=0} = 0 \qquad (8.29)$$

At 0 K, the entropy of a substance is independent of its volume and pressure. In the third law of thermodynamics, there is therefore no need to indicate the system pressure.

8.9.3 Helmholtz Energy and Gibbs Energy at 0 K

Since S is zero at 0 K, using the relations 5.23, we find that in the vicinity of 0 K :

$$-S = \left(\frac{\partial A}{\partial T}\right)_V = \left(\frac{\partial G}{\partial T}\right)_P \to 0 \tag{8.30}$$

which imply that the Helmholtz energy and Gibbs energy do not vary with temperature in the vicinity of 0 K. Using Eq. 5.24 leads to :

$$\left.\begin{array}{l} U = A - T\left(\dfrac{\partial A}{\partial T}\right)_V \\[2mm] H = G - T\left(\dfrac{\partial G}{\partial T}\right)_P \end{array}\right\} \Rightarrow \begin{cases} U(0\ K) = A(0\ K) \\ H(0\ K) = G(0\ K) \end{cases} \tag{8.31}$$

Near 0 K, the Helmholtz energy tends towards the internal energy and the Gibbs energy tends towards the enthalpy. We can take the derivatives of Eq. 8.31 at constant volume or pressure respectively to obtain :

$$\left.\begin{array}{l} \left(\dfrac{\partial U}{\partial T}\right)_V = -T\left(\dfrac{\partial^2 A}{\partial T^2}\right)_V \\[2mm] \left(\dfrac{\partial H}{\partial T}\right)_P = -T\left(\dfrac{\partial^2 G}{\partial T^2}\right)_P \end{array}\right\} \Rightarrow \begin{cases} \left(\dfrac{\partial U}{\partial T}\right)_V \bigg|_{T=0} = 0 \\[2mm] \left(\dfrac{\partial H}{\partial T}\right)_P \bigg|_{T=0} = 0 \end{cases} \tag{8.32}$$

If we assume that the second derivatives of A and G are finite, we find that U (at constant volume) and H (at constant pressure) are independent of temperature near 0 K.

9. Energetics of Chemical Reactions

9.1 Introduction

Consider a *closed system* in which a chemical reaction can take place. The reaction is written as:

$$|v_A| A + |v_B| B \longrightarrow |v_C| C + |v_D| D \qquad (9.1)$$

The $|v|$ are the stoichiometric coefficients of the reaction. They are *positive integers* or *simple fractional numbers*. The stoichiometric coefficients indicate that matter is conserved in a chemical reaction. The numbers of the various atoms in the reactants (A, B) are the same as in the products (C, D). We can also write reaction 9.1 as:

$$v_C C + v_D D + v_A A + v_B B = 0 \qquad (9.2)$$

where the stoichiometric coefficients are now algebraic. The coefficients of the products C and D are conventionally taken as *positive* while those of the reactants A and B are *negative*. A general reaction can be written:

$$\sum_i v_i M_i = 0 \qquad i = 1, 2, ..., n \qquad (9.3)$$

where the M_i refer to chemical species i, and n is the number of species present in the system. If a species present in the system does not take part in the reaction, its stoichiometric coefficient is simply zero. When several reactions may take place simultaneously, they are represented by:

$$\sum_i v_{i,k} M_i = 0 \qquad \begin{cases} i = 1, 2, ..., n \\ k = 1, 2, ..., r \end{cases} \qquad (9.4)$$

Each of the r reactions written in this way implies that the atoms in the chemical species are conserved during a chemical reaction.

9.2 The Extent of Reaction

Consider a *closed system* where only one reaction can take place. The variations of the number of moles of each species, n_i, are not independent. We have:

$$\frac{dn_1}{v_1} = \ldots = \frac{dn_i}{v_i} = \ldots = d\xi \tag{9.5}$$

where the variable ξ (units of mol) relates the changes in the amount of the chemical species present. By integrating the system of differential Eq. 9.5 and taking $\xi = 0$ as the initial state of the system, we get:

$$n_1 = n_1^0 + v_1 \xi \quad \ldots \quad n_i = n_i^0 + v_i \xi \quad \ldots \tag{9.6}$$

Since the numbers of moles cannot be negative, the range of valid ξ values is limited. ξ is called the **extent of the reaction** and is an extensive variable.

Example

The reaction between hydrogen and iodine to form hydrogen iodide can be represented as:

$$H_2 \text{ (g)} + I_2 \text{ (g)} \rightleftharpoons 2 \, HI \text{ (g)} \qquad 2 \, HI \text{ (g)} - H_2 \text{ (g)} - I_2 \text{ (g)} = 0$$

Consider a system containing initially 0.3 mole of H_2, 0.2 mole of I_2 and 0.1 mole of HI. According to value of the reaction extent, the number of moles present in the system are:

$$n_{H_2} = 0.3 - \xi \quad n_{I_2} = 0.2 - \xi \quad n_{HI} = 0.1 + 2\xi$$

The possible values of ξ are limited. ξ_{max} corresponds to the total consumption of I_2, $\xi_{max} = 0.2$ mol. ξ_{min} corresponds to the total consumption of HI, $\xi_{min} = -0.05$ mol.

9.3 Variables of Reaction

9.3.1 Gibbs Energy of Reaction

We consider a *closed system*, in which only a *single chemical reaction* can take place. Let us write the expression of the differential of its Gibbs energy. Using 5.22 and 9.5, we have:

$$\begin{aligned} dG &= V\,dp - S\,dT + \sum_i \mu_i \, dn_i = V\,dp - S\,dT + \sum_i \mu_i \, v_i \, d\xi \\ &= V\,dp - S\,dT + \Delta_r G \, d\xi \quad \text{where} \quad \Delta_r G = \sum_i v_i \mu_i \end{aligned}$$

(9.7)

The quantity $\Delta_r G$ is the **Gibbs energy of reaction**. It is an intensive variable. It depends on the composition of the system. We also have the relation:

$$\Delta_r G = \sum_i v_i \mu_i = \left(\frac{\partial G}{\partial \xi} \right)_{p,T} \tag{9.8}$$

The extensive variable ξ is associated to the intensive variable $\Delta_r G$.

9.3.2 Other Variables of Reaction

Even though many of them seem to be seldom used, other variables of reaction prove extremely valuable. Any extensive variable, X, can be considered as a function of p, T and the n_i, the numbers of moles of each species of the system. Using Eq. 9.5 and 6.22, we have:

$$\left. \begin{array}{l} dX = \left(\dfrac{\partial X}{\partial p}\right)_{T, n_i} dp + \left(\dfrac{\partial X}{\partial T}\right)_{p, n_i} dT + \sum_i \nu_i \overline{X_i}\, d\xi \\[6pt] = \left(\dfrac{\partial X}{\partial p}\right)_{T, n_i} dp + \left(\dfrac{\partial X}{\partial T}\right)_{p, n_i} dT + \Delta_r X\, d\xi \\[6pt] \text{with } \Delta_r X = \sum_i \nu_i \overline{X_i} = \left(\dfrac{\partial X}{\partial \xi}\right)_{p, T} \end{array} \right\} \quad (9.9)$$

The variables of reaction have therefore simple expressions in terms of the corresponding partial molar quantities. We have:

$$\left. \begin{array}{l} \Delta_r U = \sum_i \nu_i \overline{U_i} \quad \Delta_r S = \sum_i \nu_i \overline{S_i} \quad \Delta_r V = \sum_i \nu_i \overline{V_i} \\[6pt] \Delta_r H = \sum_i \nu_i \overline{H_i} \quad \Delta_r A = \sum_i \nu_i \overline{A_i} \quad \Delta_r C_p = \sum_i \nu_i \overline{C_{p\,i}} \end{array} \right\} \quad (9.10)$$

The variables of reaction have many useful applications.

9.3.3 Standard Variables of Reaction

A **standard variable of reaction** corresponds to the change in the considered variable for a system where the initial state contains the number of moles of reactants in the stoichiometric equation and as a final state, the number of moles of products present in the stoichiometric equation. All species, in the initial and final states, are pure and in their standard state.

- The standard state pressure $p^{\ominus}$ selected has always a value 1. It can be 1 atm (old convention) or 1 bar (most recent convention).
- The selected temperature is often 298.15 K, which corresponds to 25°C, but one can find standard state data for other temperatures.
- The standard state for a species that is gaseous under standard pressure at the temperature of interest, corresponds

to the pure substance behaving as an ideal gas at the standard pressure.
- The standard state for a solvent, a pure solid or a liquid, corresponds to the pure substance at the standard pressure $p^{\ominus}$.
- For solutes, several different standard states can be used. The various possible choices are presented in chapter 12.

The expressions for standard variables of reaction are :

$$\left. \begin{array}{ll} \Delta_r U^{\ominus} = \sum_i \nu_i U_i^{\ominus} & \Delta_r A^{\ominus} = \sum_i \nu_i A_i^{\ominus} \\ \Delta_r H^{\ominus} = \sum_i \nu_i H_i^{\ominus} & \Delta_r G^{\ominus} = \sum_i \nu_i \mu_i^{\ominus} \\ \Delta_r S^{\ominus} = \sum_i \nu_i S_i^{\ominus} & \Delta_r C_p^{\ominus} = \sum_i \nu_i C_{p\,i}^{\ominus} \\ \Delta_r V^{\ominus} = \sum_i \nu_i V_i^{\ominus} & \end{array} \right\} \quad (9.11)$$

Standard molar entropy, standard molar heat capacity and standard molar volume are intensive variables that are known in an absolute way. The other standard molar quantities in Eq. 9.11 have values that depend on the selection of a reference for the energy scale. For reaction (9.1) taking place between gaseous species, the standard variables of reaction are evaluated according to the scheme presented in table 9.1.

Initial state		Final state
$p = 1$ atm or 1 bar, T $\|\nu_A\|$ moles of pure A ideal gas behavior	Mixing Chemical process Separation $\longrightarrow$	$p = 1$ atm or 1 bar, T $\|\nu_C\|$ moles of pure C ideal gas behavior
$p = 1$ atm or 1 bar, T $\|\nu_B\|$ moles of pure B ideal gas behavior		$p = 1$ atm or 1 bar, T $\|\nu_D\|$ moles of pure D ideal gas behavior

Table 9.1 Chemical process in a *gaseous system* from an initial standard state of the reactants to a final standard state of the products, for reaction 9.1.

Example
Consider the reaction at 298.15 K :

9 Energetics of Chemical Reactions

$$H_2 \text{ (g)} + \frac{1}{2} O_2 \text{ (g)} \rightleftarrows H_2O \text{(l)}$$

We find in thermodynamic tables the standard enthalpy and the standard Gibbs energy of formation (see 9.3.4) of H_2O(l), which are also the standard enthalpy and the Gibbs energy of this reaction since the standard variables of formation of H_2 and O_2 are zero :

$$\Delta_r H^\ominus = \Delta_f H^\ominus = -285830 \text{ J mol}^{-1} \qquad \Delta_r G^\ominus = -\Delta_f G^\ominus = -237129 \text{ J mol}^{-1}$$

Note that usually the units in the tables are kJ mol^{-1}. The initial state is pure hydrogen and pure oxygen in their standard state, and the final state is pure liquid water under a 1 bar pressure.

9.3.4 Standard Variables of Formation

Standard variables of formation are usually found in chemical thermodynamic data tables.
A standard variable of formation, at a given temperature, is a standard variable of reaction that corresponds to the change in the corresponding extensive variable when **one mole of a chemical species is formed** in its *standard state* from the appropriate number of moles of the most stable state of each element it contains taken in their standard state. The standard variables of formation of an element in its most stable state at temperature T and standard state pressure $p^\ominus$ are zero, by definition. The standard variables of formation of a species at a temperature T are represented by : $\Delta_f G_T^\ominus$, $\Delta_f H_T^\ominus$, $\Delta_f S_T^\ominus$, ...
For the formation of ammonia under 1 bar at 298.15 K, we have :

$$\frac{1}{2} N_2 \text{ (g)} + \frac{3}{2} H_2 \text{ (g)} \rightleftarrows NH_3 \text{ (g)} \qquad (9.12)$$

The standard enthalpy of formation of ammonia that is found in the tables, corresponds to (Eq. 9.11) :

$$\Delta_f H^\ominus_{298.15}(NH_3) = H^\ominus_{298.15}(NH_3) - \frac{1}{2} H^\ominus_{298.15}(N_2) - \frac{3}{2} H^\ominus_{298.15}(H_2)$$
$$(9.13)$$

In thermodynamic tables, the following types of data are often presented.

$\Delta_f G^\ominus_{298.15}$	Standard Gibbs energy of formation
$\Delta_f H^\ominus_{298.15}$	Standard enthalpy of formation
$S^\ominus_{298.15}$	Standard molar entropy
$C^\ominus_{p\,298.15}$	Standard molar heat capacity at constant pressure

The first two are *standard variables of formation*. They refer to the formation of one mole of the species. The other two are *standard quantities*. The standard Gibbs energies of formation allow the calculations of the standard Gibbs energy of a reaction, $\Delta_r G^\ominus_{298.15}$, which leads to the knowledge of the equilibrium constant of a reaction (See chapter 10).

9.4 Hess' Law

Hess' Law is :

> The enthalpy of a reaction is equal to the sum of the enthalpies of other reactions into which it can be formally decomposed.

A similar statement can be formulated for any standard variable of reaction. Any standard variable of reaction can be obtained by an appropriate combination of standard variables of formation.

$$\Delta_r X^\ominus_T = \sum_i \nu_i X^\ominus_{i,T} = \sum_i \nu_i \Delta_f X^\ominus_{i,T} \tag{9.14}$$

We can write the following relations for some of the standard variables of reaction :

$$\left. \begin{array}{l} \Delta_r G^\ominus_T = \sum_i \nu_i \mu^\ominus_{i,T} = \sum_i \nu_i \Delta_f G^\ominus_{i,T} \\[6pt] \Delta_r H^\ominus_T = \sum_i \nu_i H^\ominus_{i,T} = \sum_i \nu_i \Delta_f H^\ominus_{i,T} \\[6pt] \Delta_r U^\ominus_T = \sum_i \nu_i U^\ominus_{i,T} = \sum_i \nu_i \Delta_f U^\ominus_{i,T} \\[6pt] \Delta_r S^\ominus_T = \sum_i \nu_i S^\ominus_{i,T} = \sum_i \nu_i \Delta_f S^\ominus_{i,T} \end{array} \right\} \tag{9.15}$$

Example

Consider the reaction :

$$C_2H_5OH\ (l) + 3\ O_2\ (g) \rightleftharpoons 2\ CO_2\ (g) + 3\ H_2O\ (l)$$

We use the value provided in the table below to obtain numerical values for the various standard quantities.

9 Energetics of Chemical Reactions

Species	$\Delta_f H^\ominus_{298.15}$	$S^\ominus_{298.15}$	$C^\ominus_{p\,298.15}$
Units	kJ mol^{-1}	J mol^{-1}K^{-1}	J mol^{-1}K^{-1}
C$_2$H$_5$OH (l)	-277.69	160.67	111.46
O$_2$ (g)		205.146	29.36
CO$_2$ (g)	-393.51	213.75	37.11
H$_2$O (l)	-285.83	69.91	75.291

Standard thermodynamic data (1 bar) for the oxidation of ethanol by oxygen.

The expressions for the standard enthalpy of this reaction is:

$$\Delta_r H^\ominus_{298.15} = \sum_i \nu_i \Delta_f H^\ominus_{i\,298.15}$$

$$= 2\,\Delta_f H^\ominus_{298.15}(CO_2(g)) + 3\,\Delta_f H^\ominus_{298.15}(H_2O(l))$$

$$- \Delta_f H^\ominus_{298.15}(C_2H_5OH(l))$$

$$= 2 \cdot (-393.51) + 3 \cdot (-285.83) - (-277.69)$$

$$= -1366.82 \text{ kJ mol}^{-1}$$

The standard enthalpy of reaction obtained is negative. When the reaction takes place (under standard conditions), the system delivers heat to the surroundings and such a reaction is known as **exothermic**. If the standard enthalpy of a reaction is positive, then the reaction is **endothermic**.

The standard entropy of reaction is:

$$\Delta_r S^\ominus_{298.15} = \sum_i \nu_i S^\ominus_{i\,298.15}$$

$$= 2\,S^\ominus_{298.15}(CO_2(g)) + 3\,S^\ominus_{298.15}(H_2O(l))$$

$$- S^\ominus_{298.15}(C_2H_5OH(l)) - 3\,S^\ominus_{298.15}(O_2(g))$$

$$= 2 \cdot (213.75) + 3 \cdot (69.91) - 160.67 - 3 \cdot (205.146)$$

$$= -138.88 \text{ J K}^{-1}\text{mol}^{-1}$$

The standard Gibbs energy of the reaction is:

$$\Delta_r G^\ominus_{298.15} = \Delta_r H^\ominus_{298.15} - 298.15\,\Delta_r S^\ominus_{298.15}$$

$$= -1366.82 \text{ kJ mol}^{-1} - 298.15 \text{ K} \cdot (-138.88 \text{ J K}^{-1}\text{mol}^{-1})$$

$$= -1325.41 \text{ kJ mol}^{-1}$$

9.5 Kirchhoff's Equation

It describes the effect of the temperature of a system on its enthalpy of reaction. Consider a *closed system*, in which a *single chemical reaction* can take place. We use $T, p, ..., n_i,...$ as the variables. The change of the enthalpy of the system during an infinitesimal change is obtained applying 9.9 to H:

$$dH = \left(\frac{\partial H}{\partial T}\right)_{p, n_i} dT + \left(\frac{\partial H}{\partial p}\right)_{T, n_i} dp + \sum_i \nu_i \overline{H_i}\, d\xi \qquad (9.16)$$

Which can be written using 5.43:

$$\left.\begin{aligned} dH &= C_p\, dT + V(1 - \alpha T)\, dp + \sum_i \nu_i \overline{H_i}\, d\xi \\ &= C_p\, dT + V(1 - \alpha T)\, dp + \Delta_r H\, d\xi \end{aligned}\right\} \qquad (9.17)$$

Using Schwarz theorem, we find:

$$\left(\frac{\partial \Delta_r H}{\partial T}\right)_{p, \xi} = \left(\frac{\partial C_p}{\partial \xi}\right)_{p, T} = \Delta_r C_p \qquad (9.18)$$

Using 6.20 we can express the right hand side of Eq. 9.18.

$$C_p = \sum_i n_i \overline{C_{pi}} \Rightarrow \left\{\begin{aligned} \left(\frac{\partial C_p}{\partial \xi}\right)_{p, T} &= \sum_i \overline{C_{pi}} \left(\frac{\partial n_i}{\partial \xi}\right)_{p, T} \\ \Delta_r C_p &= \sum_i \nu_i \overline{C_{pi}} \end{aligned}\right. \qquad (9.19)$$

The change of the enthalpy of reaction with temperature is:

$$\left(\frac{\partial \Delta_r H}{\partial T}\right)_{p, \xi} = \Delta_r C_p = \sum_i \nu_i \overline{C_{pi}} \qquad (9.20)$$

This equation is **Kirchhoff's equation**. This result can be applied to the standard enthalpy of reaction, which does not depend on either the pressure (which is standard) or the extent of reaction (pure substances). We have therefore:

$$\left(\frac{\partial \Delta_r H^\ominus_T}{\partial T}\right)_{p, \xi} = \sum_i \nu_i\, C^\ominus_{pi\, T} = \Delta_r C^\ominus_{p\, T} = \frac{d\Delta_r H^\ominus_T}{dT} \qquad (9.21)$$

9 Energetics of Chemical Reactions

The literature provides expressions for the molar heat capacities as developments in powers of T. Typically:

$$C^\ominus_{pi\,T} = A_i + B_i\,T + C_i\,T^2 + \ldots \qquad (9.22)$$

Example

Consider the reaction:

$$H_2 + CO_2 \rightleftharpoons CO + H_2O\ (vap)$$

For which we have the standard values:

Species	$\Delta_f H^\ominus_{298.15}$ [kJ mol^{-1}]	$S^\ominus_{298.15}$ [J mol^{-1} K^{-1}]	$C^\ominus_{p,\,298.15}$ [J mol^{-1} K^{-1}]
H_2 (g)	0	130.684	28.824
CO_2 (g)	-393.509	213.74	37.11
CO (g)	-110.525	197.674	29.142
H_2O (g)	-241.818	188.825	33.577

We can calculate standard variables of reaction at 298.15 K. We have:

$$\left.\begin{array}{l}
\Delta_r H^\ominus_{298.15} = \sum_i \nu_i \Delta_f H^\ominus_{i,\,298.15} \\[4pt]
\qquad = 1\cdot(-241.818) + 1\cdot(-110.525) - 1\cdot 0 - 1\cdot(-393.509) \\[4pt]
\qquad = 41.166\ \text{kJ mol}^{-1} \simeq 41.2\ \text{kJ mol}^{-1} \\[8pt]
\Delta_r S^\ominus_{298.15} = \sum_i \nu_i S^\ominus_{i,\,298.15} \\[4pt]
\qquad = 188.825 + 197.674 - 130.684 - 213.74 \\[4pt]
\qquad = 42.075\ \text{J mol}^{-1}\,\text{K}^{-1} \simeq 42.1\ \text{J mol}^{-1}\,\text{K}^{-1} \\[8pt]
\Delta_r G^\ominus_{298.15} = \Delta_r H^\ominus_{298.15} - T\,\Delta_r S^\ominus_{298.15} = 41.166 - 298.15\cdot(42.075\cdot 10^{-3}) \\[4pt]
\qquad = 28.621\ \text{kJ mol}^{-1} \simeq 28.6\ \text{kJ mol}^{-1} \\[8pt]
\Delta_r C^\ominus_{p\,298.15} = \sum_i \nu_i C^\ominus_{p\,i\,298.15} \\[4pt]
\qquad = (33.577 + 29.142 - 37.11 - 28.824) \\[4pt]
\qquad = -3.215\ \text{J mol}^{-1}\,\text{K}^{-1} \simeq -3.22\ \text{J mol}^{-1}\,\text{K}^{-1}
\end{array}\right\}$$

Assuming that the heat capacities are constant, the change of the standard enthalpy of reaction (units J mol^{-1}) with temperature is obtained using 9.21 (and plotted in the next example):

$$\left.\begin{array}{l}
\Delta_r H^\ominus_T = \Delta_r H^\ominus_{298.15} + \Delta_r C^\ominus_{p\,298.15}(T - 298.15) \\[4pt]
\qquad = 41166 - 3.215\,(T - 298.15)
\end{array}\right\}$$

9.6 Effect of Temperature on the Entropy of Reaction and the Gibbs Energy of Reaction

Consider a *closed system*, in which a *single chemical reaction* can take place. Using $T, p, ..., n_i, ...$ as the variables, the change in entropy of the system during an infinitesimal change is obtained applying 9.9 to S:

$$dS = \left(\frac{\partial S}{\partial T}\right)_{p, n_i} dT + \left(\frac{\partial S}{\partial p}\right)_{T, n_i} dp + \sum_i \nu_i \overline{S_i} \, d\xi \tag{9.23}$$

which can be written:

$$\left.\begin{array}{l} dS = \dfrac{C_p}{T} dT - \alpha V dp + \sum_i \nu_i \overline{S_i} \, d\xi \\[2mm] = \dfrac{C_p}{T} dT - \alpha V dp + \Delta_r S \, d\xi \end{array}\right\} \tag{9.24}$$

Using Schwarz theorem and the result in Eq. 9.24 and 9.20, we find:

$$\left.\begin{array}{l} \left(\dfrac{\partial \Delta_r S}{\partial T}\right)_{p, \xi} = \left(\dfrac{\partial \frac{C_p}{T}}{\partial \xi}\right)_{p, T} = \dfrac{1}{T}\left(\dfrac{\partial C_p}{\partial \xi}\right)_{p, T} = \dfrac{1}{T} \sum_i \nu_i \overline{C_{p\,i}} \\[3mm] \left(\dfrac{\partial \Delta_r S}{\partial T}\right)_{p, \xi} = \dfrac{1}{T}\left(\dfrac{\partial \Delta_r H}{\partial T}\right)_{p, \xi} \end{array}\right\} \tag{9.25}$$

The differential expression of the Gibbs energy (Eq. 5.22) expressed as a function of p, T, ξ, leads to:

$$\left.\begin{array}{l} dG = V dp - S dT + \sum_i \mu_i \, dn_i \\[2mm] = V dp - S dT + \sum_i \nu_i \mu_i \, d\xi \\[2mm] = V dp - S dT + \Delta_r G \, d\xi \end{array}\right\} \Rightarrow \left(\frac{\partial \Delta_r G}{\partial T}\right)_{p, \xi} = -\left(\frac{\partial S}{\partial \xi}\right)_{p, T} = -\Delta_r S$$

$$\tag{9.26}$$

The results of 9.25 and 9.26 can be applied to standard variables of reaction to obtain:

$$\frac{d\Delta_r S^\ominus}{dT} = \frac{1}{T} \Delta_r C_p^\ominus \qquad \frac{d\Delta_r G^\ominus}{dT} = -\Delta_r S^\ominus \tag{9.27}$$

9 Energetics of Chemical Reactions

Example

The standard Gibbs energy as a function of temperature is given by:

$$\Delta_r G_T^\ominus = \Delta_r H_T^\ominus - T \Delta_r S_T^\ominus$$

For the reaction of the example in § 9.5, we already obtained $\Delta_r H_T^\ominus$. Assuming again that the heat capacities are constant (independent of temperature), we can evaluate the standard entropy of the reaction as a function of temperature. We can then obtain the standard Gibbs energy of reaction. We give a graphic representation of these standard quantities.

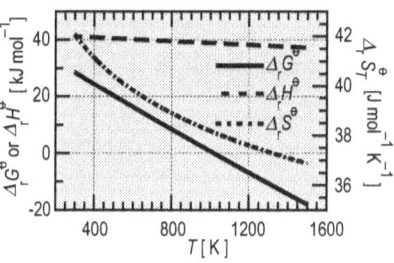

$$\frac{d \Delta_r S_T^\ominus}{dT} = \frac{\Delta_r C_p^\ominus T}{T} \Rightarrow \Delta_r S_T^\ominus = \Delta_r S_{298.15}^\ominus + \int_{298.15}^{T} \frac{\Delta_r C_p^\ominus}{T}{}_{298.15} dT$$

$$\Delta_r S_T^\ominus = \Delta_r S_{298.15}^\ominus + \Delta_r C_p^\ominus{}_{298.15} \ln\left(\frac{T}{298.15}\right)$$

$$\Delta_r S_T^\ominus = 42.075 - 3.215 \ln\left(\frac{T}{298.15}\right) \quad \text{J mol}^{-1}\text{K}^{-1}$$

9.7 Conversion of Chemical Energy into Work

9.7.1 Any Form of Work

We consider an infinitesimal *isothermal and isobaric change* of a *closed system* where a *single chemical reaction* can take place. The corresponding Helmholtz energy change is obtained applying 9.9 to A:

$$dA = \left(\frac{\partial A}{\partial p}\right)_{T, n_i} dp + \left(\frac{\partial A}{\partial T}\right)_{p, n_i} dT + \sum_i \overline{v_i A_i} \, d\xi \quad (9.28)$$

$$= \Delta_r A \, d\xi \quad \text{at constant } T \text{ and } p$$

Using Eq. 5.67, the work done on the system in any form when the extent of reaction changes by $d\xi$ is:

$$dw = dA + T dS_{\text{global}}$$
$$= \Delta_r A \, d\xi + T dS_{\text{global}} \quad (9.29)$$
$$dw \geq \Delta_r A \, d\xi$$

The Helmholtz energy of reaction (when negative), provides some quantitative information on the (maximum) amount of work (*under any form*) that a user may obtain from a chemical reaction.

9.7.2 Work other than Work due to Volume Change

We consider an infinitesimal *isothermal and isobaric change* of a *closed system* where a *single chemical reaction* can take place. The work done on the system (other than work due to volume change) is related to the change in the Gibbs energy. Using Eq. 5.73 and 9.7, we have:

$$\begin{aligned} dw_{other} &= dG + T\,dS_{global} \\ &= \Delta_r G\,d\xi + T\,dS_{global} \\ dw_{other} &\geq \Delta_r G\,d\xi \end{aligned} \quad\quad (9.35)$$

The Gibbs energy of reaction (when negative), provides some quantitative information on the (maximum) amount of work (*other than work due to volume change*) that a user may obtain from a chemical reaction.

Example

The following reaction takes place in fuel cells at room temperature when an appropriate catalyst is used:

$$H_2(g) + \frac{1}{2}O_2(g) \longrightarrow H_2O\,(l)$$

	$\Delta_f H^{\ominus}_{298.15}$ (kJ mol^{-1})	$S^{\ominus}_{298.15}$ (J mol^{-1} K^{-1})
H_2 (g)	0	130.684
O_2 (g)	0	205.138
H_2O (l)	-285.83	69.91

Using the values provided in the table, we find the following standard values for this reaction:

$$\Delta_r H^{\ominus}_{298.15} = \sum_i \nu_i \Delta_f H^{\ominus}_{i,\,298.15} = -285.83 \text{ kJ mol}^{-1}$$

$$\Delta_r S^{\ominus}_{298.15} = \sum_i \nu_i S^{\ominus}_{i,\,298.15} = 1 \cdot 69.91 + (-1)\cdot 130.684 + (-\tfrac{1}{2})\cdot 205.138$$

$$= -163.34 \text{ J mol}^{-1}\text{ K}^{-1}$$

$$\Delta_r G^{\ominus}_{298.15} = \Delta_r H^{\ominus}_{298.15} - T\Delta_r S^{\ominus}_{298.15} = -285.83 - 298.15 \cdot (-163.34)\,10^{-3}$$

$$= -237.129 \text{ kJ mol}^{-1} \simeq -237.1 \text{ kJ mol}^{-1}$$

9 Energetics of Chemical Reactions

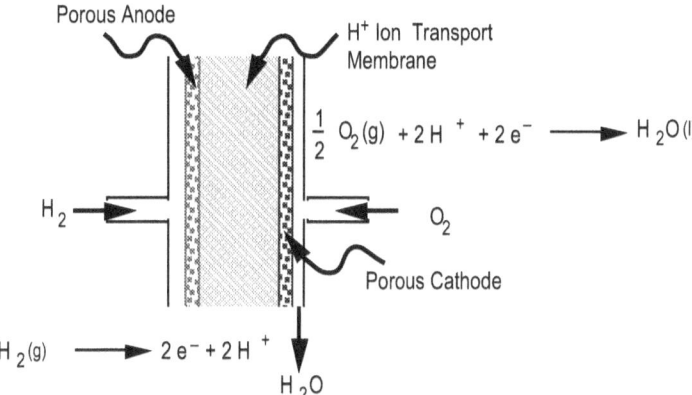

Hydrogen and oxygen are consumed to produce water. The work w_{other} that can be obtained when one mole of water is formed under standard conditions is 237 kJ. The electrochemical reactions that take place at the electrodes are :

Anode $H_2(g) \longrightarrow 2e^- + 2H^+$

Cathode $\frac{1}{2} O_2(g) + 2H^+ + 2e^- \longrightarrow H_2O(l)$

Two electrons ($n = 2$) are exchanged during the reaction. The standard potential of the cell is given by :

$$E^\ominus_{298.15} = -\frac{\Delta_r G^\ominus_{298.15}}{nF} = \frac{237.129 \cdot 10^3}{2 \cdot 9.6485 \cdot 10^4} = 1.2288 \text{ V} \simeq 1.229 \text{ V}$$

10. Chemical Equilibria

10.1 Change in $G(\xi)$ with the Extent of Reaction

10.1.1 Expression for a Mixture of Reacting Ideal Gases

Let us illustrate the relation between the Gibbs energy of reaction and the extent of reaction. Consider a *closed isothermal and isobaric system* where all chemical species are ideal gases and reaction 9.3 is the only reaction that can take place. Assume that only *work of volume* can be done on the system. Prior to any reaction, the initial composition of the system is:

$$\text{Initial composition} \quad n_i^0 \text{ moles of species } i \quad i = 1, 2, \ldots, n \quad (10.1)$$

Due to the stoichiometry of the reaction, the changes in the number of moles of each species are related to changes in ξ. We have seen that:

$$\frac{dn_1}{v_1} = \ldots = \frac{dn_i}{v_i} = \ldots = d\xi \quad (9.5)$$

In the initial state of the system, no reaction has yet taken place and $\xi = 0$. For a reaction extent ξ, the number of moles of each of the species is given by:

$$n_1 = n_1^0 + v_1 \xi \quad \ldots \quad n_i = n_i^0 + v_i \xi \quad \ldots \quad (9.6)$$

A positive value of ξ corresponds to the reaction proceeding to the right and a negative value of ξ corresponds to the reaction proceeding to the left. As already mentioned in chapter 9, the allowed values of ξ are restricted to limits corresponding to the total disappearance of either one of the reactants or one of the products. The maximum and minimum possible value for ξ are:

$$\left. \begin{array}{l} \xi_{min} = \max\left(\ldots, -\dfrac{n_i^0}{v_i}, \ldots\right) \text{ for the positive } v_i \\[1em] \xi_{max} = \min\left(\ldots, -\dfrac{n_i^0}{v_i}, \ldots\right) \text{ for the negative } v_i \end{array} \right\} \quad (10.2)$$

Using the expression of the chemical potential of ideal gases, (Eq. 7.16) and the expression of the Gibbs energy (Eq. 6.16), we obtain the expression for the Gibbs energy of the system as a function of ξ:

$$\mu_i = \mu_i^\ominus(T) + RT\ln p + RT\ln x_i \qquad G = \sum_i n_i \mu_i$$

$$x_i = \frac{n_i}{\sum_j n_j} = \frac{n_i^0 + \nu_i \xi}{\sum_j (n_j^0 + \nu_j \xi)} \qquad (10.3)$$

and finally :

$$G(\xi) = \sum_i (n_i^0 + \nu_i \xi)\left(\mu_i^\ominus(T) + RT\ln\left[\frac{n_i^0 + \nu_i \xi}{\sum_j (n_j^0 + \nu_j \xi)} p\right]\right) \qquad (10.4)$$

10.1.2 Schematic Representation

The function $G(\xi)$ depends on pressure, temperature, stoichiometric coefficients of the reaction, chemical potential of the species in their standard state and on the extent of reaction. We have given in Fig. 10.1 a schematic representation of $G(\xi)$ as a function of ξ, for *isobaric and isothermal* conditions. The state of equilibrium for this system corresponds to the value of ξ for which the Gibbs energy is minimum (See 5.7). The system tends to spontaneously evolve towards this state. We can now look in more detail at the equilibrium criterion and the spontaneity of chemical reactions under various conditions.

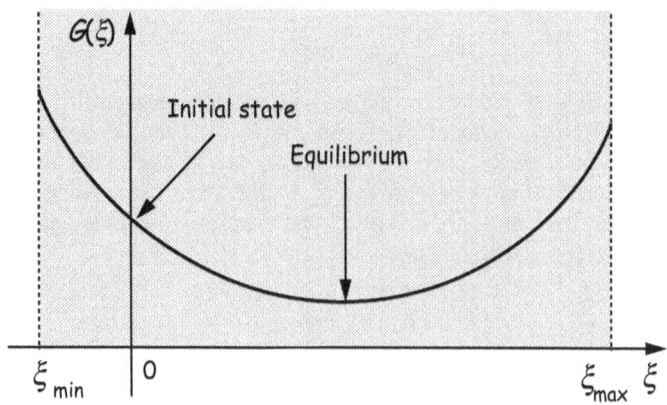

Figure 10.1 Schematic representation of $G(\xi)$ as a function of the reaction extent for *isobaric and isothermal* conditions.

10 Chemical Equilibria

10.2 Spontaneous Reaction. Equilibrium

10.2.1 Isothermal Isobaric System

Consider an *isothermal* (in contact with a thermal reservoir) *and isobaric closed system* that can only exchange work due to volume change and in which a single chemical reaction can take place. The system tends to have a spontaneous evolution such that its Gibbs energy decreases. For an infinitesimal change, $dG < 0$, (relation 5.73), Eq. 9.7 shows that :

- If $\Delta_r G > 0$, then the reaction takes place spontaneously from right to left since $dG < 0$ implies that $d\xi < 0$.
- If $\Delta_r G < 0$, then the reaction takes place spontaneously from left to right since $dG < 0$ implies that $d\xi > 0$.
- If $\Delta_r G = 0$, there is no evolution. The system is at equilibrium.

The equilibrium condition is that the Gibbs energy of reaction be zero.

$$\Delta_r G = \sum_i \nu_i \mu_i = 0 \iff \begin{cases} \text{Chemical equilibrium is reached} \\ G \text{ has reached its minimum value} \end{cases} \quad (10.5)$$

Example

Consider an isothermal isobaric (pressure p_{tot}) system where the following reaction takes place between ideal gases :

$$2\,A_{(g)} \rightleftharpoons A_2{(g)}$$

Pressures are proportional to number of moles and depend on the reaction extent. Using the properties of equal ratios, we have :

$$\left. \begin{array}{l} n_{A_2} = n^0_{A_2} + \xi \\ n_A = n^0_A - 2\xi \end{array} \right\} \Rightarrow \frac{p_{A_2}}{n^0_{A_2} + \xi} = \frac{p_A}{n^0_A - 2\xi} = \frac{p_{tot}}{n^0_A + n^0_{A_2} - \xi}$$

The value of the reaction extent ξ_{eq} that corresponds to the equilibrium is the value of ξ solution of :

$$\Delta_r G = 0 \Rightarrow \mu^\ominus_{A_2} + RT \ln \frac{n^0_{A_2} + \xi}{n^0_A + n^0_{A_2} - \xi} p_{tot} - 2\left(\mu^\ominus_A + RT \ln \frac{n^0_A - 2\xi}{n^0_A + n^0_{A_2} - \xi} p_{tot} \right) = 0$$

The equilibrium state depends on the temperature, the standard chemical potential of the species and the pressure imposed to the system. The volume at equilibrium is different from the initial volume.

$$V = \frac{(n^0_A + n^0_{A_2} - \xi_{eq})RT}{p_{tot}}$$

10.2.2 Isothermal Isochoric System

Consider an *isothermal* (in contact with a thermal reservoir) *and isochoric* (constant volume) *closed system* on which no work is done and in which a single chemical reaction can take place. The system tends to have a spontaneous evolution such that its Helmholtz energy decreases (for an infinitesimal change $\Rightarrow dA < 0$, relation 5.67). From the differential expression of dA (Eq. 5.22), we find :

$$dA = -p\,dV - S\,dT + \sum_i \mu_i\,dn_i = \sum_i \mu_i \nu_i\,d\xi$$
$$= \left(\frac{\partial A}{\partial \xi}\right)_{V,T} d\xi = \Delta_r G\,d\xi \qquad (10.6)$$

According to the sign of $\Delta_r G$, we obtain the same conclusions as in the isothermal isobaric case (in general, the system pressure changes). The equilibrium is achieved when the Gibbs energy of reaction is zero :

$$\Delta_r G = \sum_i \nu_i \mu_i = 0 \iff \begin{cases} \text{Chemical equilibrium is reached} \\ A \text{ has reached its minimum value} \end{cases} \qquad (10.7)$$

Example

Consider an isothermal isochoric system (volume V) where the following reaction takes place between ideal gases :

$$2\,A(g) \rightleftharpoons A_2(g)$$

Here again, pressures are proportional to number of moles and depend on the reaction extent. The system pressure depends also on the reaction extent.

$$\left.\begin{array}{l} n_{A_2} = n^0_{A_2} + \xi \\ n_A = n^0_A - 2\xi \end{array}\right\} \Rightarrow \frac{p_{A_2}}{n^0_{A_2} + \xi} = \frac{p_A}{n^0_A - 2\xi} = \frac{p_{tot}}{n^0_A + n^0_{A_2} - \xi} = \frac{RT}{V}$$

The value of the reaction extent ξ_{eq} that corresponds to the equilibrium is the solution of :

$$\Delta_r G = \mu^\ominus_{A_2} + RT \ln\left[(n^0_{A_2} + \xi)\frac{RT}{V}\right] - 2\left(\mu^\ominus_A + RT \ln\left[(n^0_A - 2\xi)\frac{RT}{V}\right]\right) = 0$$

The equilibrium pressure in this case depends on the volume of the system.

$$p_{eq} = (n^0_A + n^0_{A_2} - \xi_{eq})\frac{RT}{V}$$

10.2.3 Adiabatic Isobaric System

Consider an *adiabatic and isobaric closed system* that can only exchange work due to volume change and in which a single chemical reaction can take place. To determine the evolution of the system, we apply the second law. For an infinitesimal process, the heat received by the system is zero, $dq_p = 0$, the enthalpy of the system is constant. From 5.22, we obtain:

$$dH = dq_p = 0 = TdS + \sum_i \mu_i \, dn_i = TdS + \sum_i \nu_i \mu_i \, d\xi \qquad (10.8)$$

A spontaneous evolution corresponds to an increase of the system entropy, equilibrium is attained when its entropy is maximum.

$$dS \geq 0 \quad \Rightarrow \quad -\sum_i \nu_i \mu_i \, d\xi \geq 0 \quad \Rightarrow \quad \Delta_r G \, d\xi \leq 0 \qquad (10.9)$$

According to the sign of $\Delta_r G$, we obtain the same conclusions as in the isothermal isobaric case. The equilibrium condition is:

$$\Delta_r G = \sum_i \nu_i \mu_i = 0 \quad \Leftrightarrow \quad \begin{cases} \text{Chemical equilibrium is reached} \\ \text{Entropy of the closed adiabatic system} \\ \text{has reached its maximum} \end{cases} \qquad (10.10)$$

Example

In the adiabatic isobaric case, the equilibrium state is determined by:

$$\Delta_r G = 0 \quad \text{and} \quad H_F - H_I = q_p = 0$$

The reaction extent at equilibrium and the final system temperature are determined by these two equations. The second equation can be expressed in terms of partial molar quantities of the system in its initial and final states.

$$0 = \sum_i (n_i^0 + \nu_i \xi) \, \overline{H}_{iF} - \sum_i n_i^0 \, \overline{H}_{iI}$$

For a system of ideal gases, the first equation has already been written in the case of the isothermal isobaric case (Eq. 10.4). Assuming the molar heat capacities of the species are constant (thus equal to the standard heat capacities), the second equation is:

$$\left. \begin{array}{l} H_F - H_I = q_p = 0 = \sum_i (n_i^0 + \nu_i \xi) \, H_i^\ominus(T_F) - \sum_i n_i^0 \, H_i^\ominus(T_I) \\ \text{where} \quad H_i^\ominus(T_F) = H_i^\ominus(T_I) + C_{p\,i}^\ominus(T_F - T_I) \end{array} \right\}$$

Finally, we obtain an equation for the final temperature of the system also in terms of the reaction extent:

$$0 = \xi \Delta_r H^\theta(T_l) + \xi(T_F - T_l)\Delta_r C_p^\theta + \sum_i n_i^0 C_{p\,i}^\theta (T_F - T_l)$$

$$T_F = T_l - \frac{\xi \Delta_r H^\theta(T_l)}{\xi \Delta_r C_p^\theta + \sum_i n_i^0 C_{p\,i}^\theta}$$

10.2.4 Adiabatic Isochoric System

Consider an *adiabatic and isochoric closed system* (constant volume) in which a single chemical reaction can take place. No work is done on the system. To determine the evolution of the system, we apply the second law. For an infinitesimal process, the heat received by the system is zero, $dq_V = 0$, the internal energy of the system is constant. From 5.22, we obtain:

$$dU = dq_V = 0 = TdS + \sum_i \nu_i \mu_i \, d\xi \qquad (10.11)$$

A spontaneous evolution corresponds to an increase of the system entropy, equilibrium is attained when its entropy is maximum.

$$dS \geq 0 \;\Rightarrow\; -\sum_i \nu_i \mu_i \, d\xi \geq 0 \;\Rightarrow\; \Delta_r G\, d\xi \leq 0 \qquad (10.12)$$

According to the sign of $\Delta_r G$, we obtain the same conclusions as in the isothermal isobaric case. The equilibrium condition is that obtained in 10.2.3.

$$\Delta_r G = \sum_i \nu_i \mu_i = 0 \;\Leftrightarrow\; \begin{cases} \text{Chemical equilibrium is reached} \\ \text{Entropy of the closed adiabatic system} \\ \text{has reached its maximum} \end{cases}$$

$$(10.10)$$

Example

In the adiabatic isochoric case, the equilibrium state is determined by:

$\Delta_r G = 0$ and $U_F - U_l = 0$

The second equation would be expressed in terms of partial molar quantities.

$$0 = \sum_i (n_i^0 + \nu_i \, \xi) \, \overline{U}_{iF} - \sum_i n_i^0 \, \overline{U}_{il}$$

A spontaneous evolution takes place if $\Delta_r G\, d\xi \leq 0$. When the system has reached equilibrium, the Gibbs energy of reaction is zero.

10 Chemical Equilibria

10.3 Law of Mass Action for a Gas Mixture

10.3.1 Standard Equilibrium Constant

We consider a closed system at pressure p and temperature T containing an ideal gas mixture that can react according to reaction 9.3 :

$$\sum_i \nu_i M_i = 0 \qquad i = 1, 2, ..., n \qquad (10.13)$$

At equilibrium, the Gibbs energy of reaction of the system is zero. Let us assume the gases behave ideally and use the expression (Eq. 7.16) of the chemical potential for a gas in a mixture in Eq. 10.5 to obtain :

$$\left. \begin{aligned} \Delta_r G = \sum_i \nu_i \mu_i = 0 = \sum_i \nu_i \left(\mu_i^\ominus(T) + R T \ln \frac{p_i}{p^\ominus} \right) \\ 0 = \sum_i \nu_i \mu_i^\ominus(T) + R T \sum_i \nu_i \ln \frac{p_i}{p^\ominus} \\ 0 = \sum_i \nu_i \mu_i^\ominus(T) + R T \ln \prod_i \left(\frac{p_i}{p^\ominus} \right)^{\nu_i} \end{aligned} \right\} \qquad (10.14)$$

The equilibrium condition is usually written :

$$\left. \begin{aligned} \prod_i \left(\frac{p_i}{p^\ominus} \right)^{\nu_i} = K^\ominus \\ \text{with} \quad R T \ln K^\ominus = -\Delta_r G_T^\ominus = -\sum_i \nu_i \mu_i^\ominus(T) \end{aligned} \right\} \qquad (10.15)$$

The first Eq. of 10.15 is the **law of mass action**. The constant $K^\ominus$ is the **standard equilibrium constant**, it is a dimensionless quantity. However, note that the numerical value of the standard equilibrium constant depends on temperature, on the standard state pressure and on the standard state convention selected. The standard equilibrium constant is sometimes also represented by the symbol K and designated then as the **thermodynamic equilibrium constant**. If gases are not ideal, the equilibrium condition is very similar to 10.15 where partial pressures are replaced by the fugacity of the gaseous species using 7.44 to express the chemical potential.

Example

Consider the reaction, where all species behave as ideal gases:

$$H_2 + CO_2 \rightleftharpoons CO + H_2O \text{ (vap)}$$

In the examples of § 9.5 and 9.6, we obtained the standard enthalpy and entropy of reaction as a function temperature. At 800K, we obtain the following values for standard variables of reaction:

$$\Delta_r H^\ominus_{800} = 39553 \text{ J mol}^{-1} \quad \Delta_r S^\ominus_{800} = 38.902 \text{ J mol}^{-1}\text{K}^{-1} \quad \Delta_r G^\ominus_{800} = 8431 \text{ J mol}^{-1}$$

The standard equilibrium constant at this temperature is:

$$K^\ominus = \exp\left(-\frac{\Delta_r G^\ominus_{800}}{RT}\right) = \exp\left(\frac{-8431}{8.3145 \cdot 800}\right) \simeq 0.282$$

Assume that we mixed 1 mole of H_2 with 1 mole of CO_2 in a 50 liter reactor. For a reaction extent ξ, the partial pressures are:

$$p_{H_2} = p_{CO_2} = \frac{(1-\xi)RT}{V} \qquad p_{CO} = p_{H_2O} = \frac{\xi RT}{V}$$

The reaction extent at equilibrium is the solution of:

$$K^\ominus = \frac{p_{CO}\, p_{H_2O}}{p_{H_2}\, p_{CO_2}} = \frac{\xi^2}{(1-\xi)^2} \quad \Rightarrow \quad \sqrt{K^\ominus} = \frac{\xi}{1-\xi} \quad \Rightarrow \quad \xi_{eq} \simeq 0.347 \text{ mol}$$

We find the partial pressures at equilibrium:

$$p_{H_2} = p_{CO_2} = \frac{(1-\xi_{eq})RT}{V} \simeq 0.869 \text{ bar} \qquad p_{CO} = p_{H_2O} = \frac{\xi_{eq} RT}{V} \simeq 0.462 \text{ bar}$$

10.3.2 Other Forms of the Law of Mass Action

Other equilibrium constants are often used and the law of mass action can be written in different forms. The algebraic sum of the stoichiometric coefficients is conventionally represented by:

$$\Delta \nu = \sum_i \nu_i \qquad (10.16)$$

We can define an equilibrium constant K_p, relative to pressures. We have, using Eq. 10.15:

$$\left.\begin{array}{l} \prod_i p_i^{\nu_i} = K_p \qquad \text{where } K_p = K^\ominus (p^\ominus)^{\Delta \nu} \\ \\ K_p = K^\ominus \text{ numerically} \qquad \text{because } p^\ominus = 1 \text{ bar} \\ \qquad\qquad\qquad\qquad\quad \text{and pressure unit is bar} \end{array}\right\} \qquad (10.17)$$

The constant K_p is in general not dimensionless, unless $\Delta \nu$ happens to be zero for the reaction of interest. Its magnitude is also affected by the pressure unit selected to express the law of mass action. If

10 Chemical Equilibria

the law of mass action is written using the absolute values of the stoichiometric coefficients, the partial pressures of the products are in the numerator, while those of the reactants are in the denominator. Other forms of the law of mass action are encountered where different composition scales are used. We have:

$$\left. \begin{array}{l} K^\ominus = (p^\ominus)^{-\Delta v} \prod_i p_i^{v_i} = \left(\dfrac{p}{p^\ominus}\right)^{\Delta v} \prod_i x_i^{v_i} \\[2ex] = \left(\dfrac{p}{p^\ominus n}\right)^{\Delta v} \prod_i n_i^{v_i} = \left(\dfrac{RT}{p^\ominus V}\right)^{\Delta v} \prod_i n_i^{v_i} \end{array} \right\} \qquad (10.18)$$

where n is the total number of moles of gas in the system and we have made use of the ideal gas law.

Example

Consider the reaction at 450 K (note this is the same reaction as reaction 9.12 written with different stoichiometric coefficients):

$$N_2 + 3H_2 \rightleftharpoons 2NH_3$$

For this reaction, we have:

$$K^\ominus \simeq 1.4 \qquad \Delta v = \sum_i v_i = 2 - 3 - 1 = -2$$

The value $K^\ominus$ is to be used when pressures are expressed in bar. The equilibrium constant K_p depends on the pressure units used to express the law of mass action. We have:

$$\left. \begin{array}{ll} \text{pressures in bar} & K_p \simeq 1.4 \text{ bar}^{-2} \\ \text{pressures in Pa} & K_p \simeq 1.4 \cdot (10^5)^{\Delta v} = 1.4 \; 10^{-10} \text{ Pa}^{-2} \end{array} \right\}$$

10.4 Chemical Equilibrium in the Presence of Pure Condensed Phases

10.4.1 Chemical Potential of a Pure Condensed Phase

In chapter 7, we gave the expression for the chemical potential of ideal gases in ideal gas mixtures. To be able to express the condition for chemical equilibrium (Eq. 10.5, 10.6 or 10.10) for a system where condensed phases are present, the chemical potential must be known for each species present in the system. In a system where both pure condensed phases (single pure chemical species) and gas phase are present, the change in the chemical potential of the pure condensed phase with a change in pressure is often negligible compared to the change of chemical potential of the gaseous species. One can often write:

$$\mu \simeq \mu^{\ominus}(T) \qquad (10.19)$$

in systems that have condensed phases and a gas phase. If it becomes necessary to take into account the effect of pressure or composition of the condensed phase, one needs to take into account the activity, a, of the species in the phase (which has to be measured or calculated):

$$\mu = \mu^{\ominus}(T) + RT \ln a \qquad (10.20)$$

Example

Consider an incompressible pure liquid ($\kappa \simeq 0$). We obtain the chemical potential, with the pressure expressed in bar, as follows:

$$\left(\frac{\partial \mu}{\partial p}\right)_{T,n} = \left(\frac{\partial V}{\partial n}\right)_{T,p} = V_m \simeq V^{\ominus} \;\Rightarrow\; \mu = \mu^{\ominus}(T) + \int_1^p V^{\ominus} dp = \mu^{\ominus}(T) + 10^5 \, V^{\ominus} (p-1)$$

We find the activity by identifying Eq. 10.20 with this expression of the chemical potential:

$$a = \exp\left(\frac{10^5 \, V^{\ominus}(p-1)}{RT}\right) = \exp\left(\frac{10^5 \cdot 18 \, 10^{-6} \cdot 1}{8.3145 \cdot 298.15}\right) = \exp(7.26 \, 10^{-4}) = 1.0007 \simeq 1$$

Note the factor 10^5 to convert Pa to bar. The evaluation is for liquid water at 25°C, for an isothermal pressure change from 1 to 2 bar. We used $V^{\ominus} = 18 \, 10^{-6} \, \text{m}^3 \, \text{mol}^{-1}$. We see that in most cases one can assume an activity of 1, which is equivalent to taking the chemical potential of the species equal to its standard chemical potential. The corresponding change in the chemical potential is 1.8 J, very small compared to the change of the chemical potential of the gases in the system (1718 J for an ideal gas for the same pressure change).

10.4.2 Law of Mass Action for Heterogeneous Systems

We examine a system in which condensed phases are present but none of them is a solution[†]. Consider the reaction:

$$CaCO_3 \,(s) \;\rightleftharpoons\; CaO \,(s) + CO_2 \,(g) \qquad (10.21)$$
Calcite

The chemical potential of each solid is practically identical to its standard potential. When equilibrium is achieved, the chemical potential of each of the species present as a condensed phase is equal to its chemical potential in the gas phase. We have:

$$\left. \begin{array}{l} \mu_{CaCO_3\,(g)} = \mu_{CaCO_3\,(s)} \simeq \mu^{\ominus}_{CaCO_3\,(s)} \\[4pt] \mu_{CaO\,(g)} = \mu_{CaO\,(s)} \simeq \mu^{\ominus}_{CaO\,(s)} \end{array} \right\} \qquad (10.22)$$

We assume that the gas phase behaves as an ideal gas mixture. Using Eq. 10.22, we find:

[†] See chapters 11 and 12 for the thermodynamics of solutions.

10 Chemical Equilibria

$$\Delta_r G = \sum_i \nu_i \mu_i = 0 = \mu^\ominus_{CO_2} + RT\ln\frac{p_{CO_2}}{p^\ominus} + \mu^\ominus_{CaO\,(s)} - \mu^\ominus_{CaCO_3\,(s)}$$

(10.23)

For this system, the law of mass action reduces to the following expression ($p^\ominus = 1$ bar):

$$\left. \begin{array}{l} p_{CO_2} = K_p \quad \text{where} \quad K_p = K^\ominus p^\ominus \\[4pt] \text{with} \quad RT\ln K^\ominus = -\Delta_r G^\ominus_T = -\left(\mu^\ominus_{CO_2} + \mu^\ominus_{CaO\,(s)} - \mu^\ominus_{CaCO_3\,(s)}\right) \end{array} \right\}$$

(10.24)

The expression of the standard equilibrium constant and of the standard Gibbs energy *stay the same* as those for a purely gaseous system, but in the expression of the law of mass action, one must include *only the species that are exclusively present in the gas phase*.

$$\left. \begin{array}{l} \prod_j p_j^{\nu_j} = K_p \quad \text{where} \quad j \Leftrightarrow \begin{array}{l}\text{species exclusively present}\\ \text{in the gas phase}\end{array} \\[6pt] K_p = K^\ominus (p^\ominus)^{\Delta \nu} \quad \text{where} \quad \Delta \nu = \sum_j \nu_j \end{array} \right\}$$

(10.25)

All of the species involved in the reaction must be included when computing the equilibrium constant.

$$RT\ln K^\ominus = -\Delta_r G^\ominus_T = -\sum_i \nu_i \mu^\ominus_i(T) \quad \text{where } i \Leftrightarrow \begin{array}{l}\text{All of the species}\\ \text{participating}\\ \text{in the reaction}\end{array}$$

(10.26)

Example

Consider a system where 0.01 mol of $CaCO_3$ is placed in a 1 l vessel (10^{-3} m^3) that is evacuated at room temperature. From thermodynamic data, one can obtain the standard equilibrium constant, displayed as a function of temperature. For a reaction extent ξ, we have the following number of mole present in the system. We use the ideal gas law to obtain the gas pressure. Note the 10^{-5} factor to express the pressure in bar. We have :

Species	Initial number of mole	number of mole at extent ξ	Pressure [bar]
$CaCO_3$	10^{-2}	$10^{-2} - \xi$	—
CO_2	0	ξ	$10^{-5} \xi \dfrac{RT}{V}$
CaO	0	ξ	—

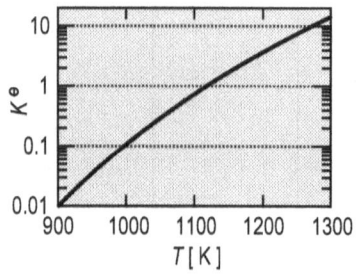
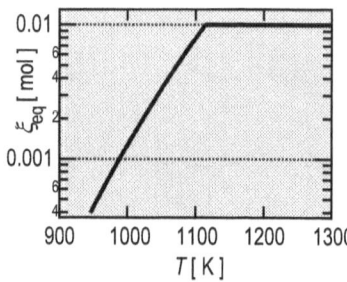

The maximum extent is $\xi_{max} = 10^{-2}$ mol. We can express the value of ξ_{eq} at some temperature using the law of mass action. We give a graphic representation of its values:

$$\frac{p_{CO_2}}{p^\ominus} = K^\ominus \implies \xi_{eq} = 10^5\, K^\ominus \frac{V}{RT}$$

We find that below 1114 K, the system is at equilibrium and all three species are present. Above that temperature, the dissociation is complete. No $CaCO_3$ is left.

10.5 Independent Reactions

10.5.1 General Remarks

An ensemble of r reactions is represented by Eq. 9.4 where the stoichiometric coefficients are positive for products and negative for reactants.

$$\sum_i \nu_{i,k}\, M_i = 0 \qquad \begin{cases} i = 1, 2, ..., n \\ k = 1, 2, ..., r \end{cases} \qquad (9.4)$$

We provide one example with which we illustrate how to find the number of reactions necessary to describe a system of reacting chemical species as well as their stoichiometric coefficients.

10.5.2 Number and Nature of Independent Reactions

Consider a system in which we know that the only species present are CO, CO_2, H_2, CH_4 and H_2O. For each of all of the possible reactions, the number of stoichiometric coefficients is equal to the number of species n (here $n = 5$). All of the possible reactions can be represented by the equation:

$$x_1\, CO + x_2\, CO_2 + x_3\, H_2 + x_4\, CH_4 + x_5\, H_2O = 0 \qquad (10.27)$$

where the ensemble of the possible coefficients x_i needs to be determined. The conservation of atoms (here H, C, O) is expressed by the following equations:

10 Chemical Equilibria

$$\begin{array}{l} H \\ C \\ O \end{array} \begin{array}{l} \\ x_1 + x_2 \\ x_1 + 2x_2 \end{array} \begin{array}{l} 2x_3 + 4x_4 + 2x_5 = 0 \\ + x_4 = 0 \\ + x_5 = 0 \end{array} \qquad (10.28)$$

If the trivial solution, where all of the coefficients x_i are zero, is the only one, it is of no interest and corresponds to a case where no chemical equilibrium between the species considered can exist. There is an infinite number of non trivial solutions $\vec{x}$ of this system, written as $A\vec{x} = \vec{0}$ if and only if the rank ρ of matrix A is smaller than the number of components of $\vec{x}$ ($n = 5$)[†]. The general solution of Equation 10.28 can be found using the *Gauss elimination method*[††]. First we write the system matrix A, then by a succession of elementary transformations, we obtain an equivalent matrix that corresponds to a simpler system of equations. We can multiply a line by a scalar different from zero and add lines to one another. If necessary, rows or columns can be exchanged in order to make all of the elements of the diagonal, starting from the first element equal to 1 (one needs to eventually remember the change of the order in the variables). The matrix A and the sequence of operations as described are the following:

$$\begin{pmatrix} 0 & 0 & 2 & 4 & 2 \\ 1 & 1 & 0 & 1 & 0 \\ 1 & 2 & 0 & 0 & 1 \end{pmatrix} \begin{array}{c} \text{exchange} \\ \text{lines} \\ \text{1 and 2} \\ \Rightarrow \end{array} \begin{pmatrix} 1 & 1 & 0 & 1 & 0 \\ 0 & 0 & 2 & 4 & 2 \\ 1 & 2 & 0 & 0 & 1 \end{pmatrix} \begin{array}{c} \text{subtract} \\ \text{line 1} \\ \text{from line 3} \\ \Rightarrow \end{array}$$

$$\begin{pmatrix} 1 & 1 & 0 & 1 & 0 \\ 0 & 0 & 2 & 4 & 2 \\ 0 & 1 & 0 & -1 & 1 \end{pmatrix} \begin{array}{c} \text{exchange} \\ \text{lines} \\ \text{2 and 3} \\ \Rightarrow \end{array} \begin{pmatrix} 1 & 1 & 0 & 1 & 0 \\ 0 & 1 & 0 & -1 & 1 \\ 0 & 0 & 2 & 4 & 2 \end{pmatrix} \begin{array}{c} \text{multiply} \\ \text{line 3} \\ \text{by } \frac{1}{2} \\ \Rightarrow \end{array}$$

$$\begin{pmatrix} 1 & 1 & 0 & 1 & 0 \\ 0 & 1 & 0 & -1 & 1 \\ 0 & 0 & 1 & 2 & 1 \end{pmatrix}$$

The rank ρ of this matrix is therefore three ($\rho = 3$). The last matrix obtained corresponds to a system equivalent to system 10.28.

[†] The rank of a matrix is, by definition, ρ, if at least one its minors of order ρ is $\neq 0$, and if all of its minors of order $\rho + 1$ are zero. A minor is a determinant formed from elements of rows and columns of the matrix.

[††] A number of mathematical programs allow to solve this system of equations.

$$\left.\begin{array}{rl} x_1 + x_2 + x_4 &= 0 \\ x_2 - x_4 + x_5 &= 0 \\ x_3 + 2\,x_4 + x_5 &= 0 \end{array}\right\} \qquad (10.29)$$

In the present case, two vectors form a basis set. We select x_4 and x_5 such that the two vectors are linearly independent. We obtain:

$$\left.\begin{array}{l} \begin{array}{l} x_4 = 1 \\ x_5 = 0 \end{array} \; \vec{x}_a = \begin{pmatrix} -2 \\ 1 \\ -2 \\ 1 \\ 0 \end{pmatrix} \;\Leftrightarrow\; -2\,CO + CO_2 - 2\,H_2 + CH_4 = 0 \\[2em] \begin{array}{l} x_4 = 0 \\ x_5 = 1 \end{array} \; \vec{x}_b = \begin{pmatrix} 1 \\ -1 \\ -1 \\ 0 \\ 1 \end{pmatrix} \;\Leftrightarrow\; CO - CO_2 - H_2 + H_2O = 0 \end{array}\right\}$$

(10.30)

Each vector corresponds to the stoichiometric coefficients of one independent reaction. The *number of independent reactions* is $r = n - \rho = 5 - 3 = 2$. Any linear combination of reactions 10.30 represents another possible reaction that can occur between the species. The reactions we obtained can of course also be written in a more conventional form:

$$\left.\begin{array}{rcl} 2\,H_2 + 2\,CO &\rightleftharpoons& CO_2 + CH_4 \\ CO_2 + H_2 &\rightleftharpoons& CO + H_2O \end{array}\right\} \qquad (10.31)$$

Example

Let us find the possible reactions in a system containing $CO_2(g)$, $CaO(s)$ and $CaCO_3(s)$. All the possible reactions can be written:

$$x_1\,CO_2(g) + x_2\,CaO(s) + x_3\,CaCO_3(s) = 0$$

Atoms are conserved during reactions and we have:

$$\left.\begin{array}{ll} C & x_1 + x_3 = 0 \\ O & 2\,x_1 + x_2 + 3\,x_3 = 0 \\ Ca & x_2 + x_3 = 0 \end{array}\right\}$$

Let us write the system matrix and transform it using Gauss' elimination.

10 Chemical Equilibria

$$\begin{pmatrix} 1 & 0 & 1 \\ 2 & 1 & 3 \\ 0 & 1 & 1 \end{pmatrix} \xrightarrow[\Rightarrow]{\substack{\text{subtract} \\ 2 \times \text{line 1} \\ \text{from line 2}}} \begin{pmatrix} 1 & 0 & 1 \\ 0 & 1 & 1 \\ 0 & 1 & 1 \end{pmatrix} \xrightarrow[\Rightarrow]{\substack{\text{subtract} \\ \text{line 2} \\ \text{from line 3}}} \begin{matrix} x_1 & x_2 & x_3 \\ \begin{pmatrix} 1 & 0 & 1 \\ 0 & 1 & 1 \\ 0 & 0 & 0 \end{pmatrix} \end{matrix}$$

The rank ρ of the matrix of this system is $\rho = 2$. The number of independent reactions is therefore $r = n - \rho = 3 - 2 = 1$. The system is equivalent to :

$$\left. \begin{matrix} x_1 & + x_3 = 0 \\ x_2 & + x_3 = 0 \end{matrix} \right\} \text{selecting } x_3 = -1 \Rightarrow \begin{cases} x_1 = 1 \\ x_2 = 1 \end{cases}$$

Hence the only possible reaction is :

$$CaCO_{3s(s)} \rightleftharpoons CO_{2(g)} + CaO_{(s)}$$

10.5.3 Equilibrium of Systems where Several Reactions can Take Place Simultaneously

In the case of a system containing n chemical species where r *independent reactions* can occur, one can write :

$$\left. \begin{matrix} \dfrac{dn_{1,k}}{\nu_{1,k}} = \ldots = \dfrac{dn_{i,k}}{\nu_{i,k}} = \ldots = d\xi_k \\ \Downarrow \\ dn_i = \sum_k dn_{i,k} = \sum_k \nu_{i,k}\, d\xi_k \end{matrix} \right\} \quad \begin{cases} i = 1, 2, \ldots, n \\ k = 1, 2, \ldots, r \end{cases} \quad (10.32)$$

where ξ_k is the extent of the kth reaction. Using the expression of dU, we find in terms of the reaction extents (5.22) :

$$\left. \begin{matrix} dU = -p\,dV + T\,dS + \sum_i \mu_i \sum_k \nu_{i,k}\, d\xi_k \\ \\ = -p\,dV + T\,dS + \sum_k \Delta_r G_k\, d\xi_k \end{matrix} \right\} \quad (10.33)$$

$\Delta_r G_k$ is the Gibbs energy of reaction k. In a similar way, we obtain the differentials of H, A and G.

$$\left. \begin{aligned} dH &= T\,dS + V\,dp + \sum_k \Delta_r G_k\, d\xi_k \\ dA &= -p\,dV - S\,dT + \sum_k \Delta_r G_k\, d\xi_k \\ dG &= V\,dp - S\,dT + \sum_k \Delta_r G_k\, d\xi_k \end{aligned} \right\} \quad (10.34)$$

For the various experimental conditions examined in 10.2, we find the condition for equilibrium or a spontaneous evolution of the system. We have:

$$\sum_k \Delta_r G_k\, d\xi_k \leq 0 \quad (10.35)$$

Equilibrium is achieved when the equality in Eq. 10.35 is true. Since the ξ_k are independent, we must have:

$$\Delta_r G_k = \sum_k \nu_{i,k}\, \mu_i = 0 \quad \text{for} \quad k = 1, 2, \ldots, r \quad (10.36)$$

At equilibrium, Eq. 10.36 provides r independent relations between the chemical potentials of the species.

Some spontaneous change can take place as long as the inequality 10.35 is true. The *essential difference* with the case of a single reaction is that some of the reactions that could definitely not take place on their own, may occur. Reactions are said to be coupled. *This type of processes takes place in all living organisms and is the essence of life.*

10.5.4 Consequences on Equilibrium

For a system at equilibrium, with r independent reactions, any reaction that is a linear combination of the independent reactions corresponds to an equilibrium ($\vec{\nu}$ = stoichiometric coefficients vector).

$$\left. \begin{aligned} \vec{\nu} &= \sum_k \lambda_k \vec{\nu}_k \quad (k = 1, \ldots, k, \ldots, r) \\ \sum_i \nu_i\, \mu_i &= \sum_i \sum_k \lambda_k \nu_{i,k}\, \mu_i = \sum_k \lambda_k \sum_i \nu_{i,k}\, \mu_i = 0 \end{aligned} \right\} \quad (10.37)$$

Hence the equilibrium of reaction $\vec{\nu}$ is achieved. The equilibrium constant of such a reaction is obtained from the equilibrium constants of the independent reactions selected to express it.

$$\ln K^\ominus = -\frac{1}{RT} \sum_i \nu_i \mu_i^\ominus(T) = -\frac{1}{RT} \sum_i \sum_k \lambda_k \nu_{i,k} \mu_i^\ominus(T)$$
$$= -\frac{1}{RT} \sum_k \lambda_k \sum_i \nu_{i,k} \mu_i^\ominus(T) = \sum_k \lambda_k \ln K_k^\ominus \qquad (10.38)$$

Example

Let us assume that we know the standard equilibrium constant for the following reactions at T:

Fe (s) + CO_2 (g) $\rightleftarrows$ FeO (s) + CO (g) $K_1^\ominus$

Fe (s) + H_2O (g) $\rightleftarrows$ FeO (s) + H_2 (g) $K_2^\ominus$

By subtracting the second reaction from the first one, we obtain:

H_2 (g) + CO_2 (g) $\rightleftarrows$ H_2O (g) + CO (g) $K_3^\ominus$

The equilibrium constant for this reaction is obtained as:

$\ln K_3^\ominus = \ln K_1^\ominus - \ln K_2^\ominus$

10.6 Phase Rule for Systems with Chemical Reactions

The phase rule as expressed in chapter 8, needs to be modified to take into account the fact that the *r independent chemical reactions* that can take place provide *r* independent relations between the chemical potentials of the species. The relations that exist between intensive variables in a system with no chemical reaction must still be satisfied. The variance decreases by *r* with respect to what it would have been if no reaction could take place. Some additional relations may exist. For example, if one of the reaction is the decomposition of a pure species, special relations may exist between the partial pressures of the decomposition products. The number of such special relations is *s*. *The phase rule for a system where chemical reactions can take place is*:

$$\left.\begin{array}{l} v = n + 2 - \varphi - r - s \\ \Downarrow \\ v = c + 2 - \varphi \quad \text{with} \quad c = n - r - s \end{array}\right\} \qquad (10.39)$$

The number *c* is the **number of components** of the system. It is the number of **independent species**. If the composition for *c* species of the system is known in one of the phases, then chemical equilibria determine the composition of all the other species in that phase.

When counting the number of species n, it is necessary to include the species that may be generated by chemical reactions.

Example

For a system containing $CO_2(g)$, $CaO(s)$ and $CaCO_3(s)$ at equilibrium, the variance is:
$v = 3 - 1 + 2 - 3 = 1$.

Example

Consider a gaseous system, hence a single phase ($\varphi = 1$), containing H_2, N_2 and NH_3. The number of chemical species of the system is $c = 3$. One can find that only one independent reaction ($r = 1$) can take place between these species. The variance of the system is $v = 3 - 1 + 2 - 1 = 3$. Selecting for example the temperature of the system and two of the partial pressures determines all of the other intensive variables. The reaction is:

$$N_2 + 3H_2 \rightleftharpoons 2NH_3$$

If in its initial state, the system was made up of pure ammonia, the stoichiometry of the reaction would imply that p_{N_2} is three times p_{H_2}, assuming that the gases behave as ideal gases.

$$p_{H_2} = 3 p_{N_2}$$

This relation between intensive variables must be taken into account and only two intensive variables can be arbitrarily fixed. The variance of the system is reduced to $v = 2$.

10.7 Effect of Temperature on the Equilibrium Constant

The effect of temperature on the equilibrium constant is obtained from Eq. 10.26. We can write:

$$\left.\begin{aligned}
\ln K^\ominus &= -\frac{\Delta_r G^\ominus_T}{RT} \\
\left(\frac{\partial \ln K^\ominus}{\partial T}\right)_p &= -\frac{1}{R}\frac{\partial}{\partial T}\left(\frac{\Delta_r G^\ominus_T}{T}\right)_p = -\frac{1}{R}\sum_i \nu_i \frac{\partial\left(\frac{\mu_i^\ominus(T)}{T}\right)_p}{\partial T} \\
&= \frac{1}{R}\sum_i \nu_i \frac{H_i^\ominus}{T^2} = \frac{1}{RT^2}\sum_i \nu_i H_i^\ominus \\
\frac{d \ln K^\ominus}{dT} &= \frac{\Delta_r H^\ominus_T}{RT^2}
\end{aligned}\right\} \quad (10.40)$$

taking into account the fact that the equilibrium constant is independent of pressure. The equation thus obtained is known as the **van't Hoff's equation**.

If the equilibrium constant is known at temperature T_I, we have:

10 Chemical Equilibria

$$\ln \frac{K^\ominus(T)}{K_I} = \int_{T_I}^{T} \frac{\Delta_r H^\ominus_T}{R T^2} dT \qquad (10.41)$$

where the expression for $\Delta_r H^\ominus_T$ is obtained using Eq. 9.21.

$$\Delta_r H^\ominus_T = \Delta_r H^\ominus_{T_I} + \int_{T_I}^{T} \Delta_r C^\ominus_{p\,T} dT \qquad (10.42)$$

Example

We know two values for the standard equilibrium constant for the reaction :

$$N_2 + 3H_2 \rightleftharpoons 2NH_3 \qquad \begin{array}{l} K^\ominus_{400} \approx 36.5 \\ K^\ominus_{450} \approx 1.39 \end{array}$$

Assuming that the enthalpy of reaction does not vary over that range, we have using Eq. 10.41 :

$$\ln \frac{K^\ominus_{T_2}}{K^\ominus_{T_1}} = \frac{\Delta_r H^\ominus}{R} \left(\frac{1}{T_1} - \frac{1}{T_2} \right) \Rightarrow \Delta_r H^\ominus = \frac{T_1 T_2}{T_2 - T_1} R \ln \frac{K^\ominus_{T_2}}{K^\ominus_{T_1}}$$

$$\Delta_r H^\ominus = \frac{400 \cdot 450}{50} \cdot 8.3145 \ln \frac{1.39}{36.5} \approx -97820 \text{ J mol}^{-1}$$

This value actually corresponds to an average value of the standard enthalpy of reaction between these two temperatures.

10.8 Displacement Laws of Equilibria

Consider a system where a single chemical reaction can take place and where equilibrium has been reached. Knowing p and T, the initial amounts of the various species present and the equilibrium constant $K^\ominus$, one can determine the equilibrium composition of the system using the law of mass action. If we modify one of the factors affecting equilibrium, it is possible to get a qualitative idea of the possible effect of this modification without prior knowledge of the equilibrium constant using the **Le Châtelier's principle**.

Le Châtelier's principle :

> Any modification of one of the factors bearing influence on the equilibrium will induce an equilibrium shift in an attempt, by the system, to prevent the imposed modification.

To describe the various effects, we assume for simplicity that the system is gaseous and is an ideal gas mixtures.

10.8.1 Effect of Temperature

According to van't Hoff's equation (10.40), we have:

$$\frac{d \ln K^\ominus}{dT} = \frac{\Delta_r H^\ominus_T}{RT^2} \tag{10.43}$$

From this equation, we can directly see how the standard equilibrium constant, $K^\ominus$, varies with temperature. We can write:

$$\left.\begin{array}{l} \text{if } \Delta_r H^\ominus_T > 0 \Rightarrow \dfrac{d \ln K^\ominus}{dT} > 0 \\[4pt] \text{endothermic reaction} \\ \Downarrow \\ \text{when } T \nearrow, K^\ominus \nearrow, \text{ equilibrium } \rightarrow \end{array}\right\} \tag{10.44}$$

$$\left.\begin{array}{l} \text{if } \Delta_r H^\ominus_T < 0 \Rightarrow \dfrac{d \ln K^\ominus}{dT} < 0 \\[4pt] \text{exothermic reaction} \\ \Downarrow \\ \text{when } T \nearrow, K^\ominus \searrow, \text{ equilibrium } \leftarrow \end{array}\right\} \tag{10.45}$$

Using the law of mass action in the form of Eq. 10.17, an increase in $K^\ominus$ (case of Eq. 10.44) implies an increase of the numerator and a concomitant decrease of the denominator. *An endothermic reaction is therefore favored by an increase in temperature. Similarly, an exothermic reaction is favored by a lowering of the temperature.*

Example

In the example of §10.7, the decrease in the equilibrium constant for a temperature raise corresponds indeed to an exothermic reaction.

10.8.2 Effect of Pressure

We can obtain the effect of pressure on the system by using the law of mass action as written in Eq. 10.18, remembering that $K^\ominus$ depends only on temperature.

$$K^\ominus = \left(\frac{p}{p^\ominus}\right)^{\Delta \nu} \prod_i x_i^{\nu_i} \tag{10.46}$$

From this expression we can conclude that:

$$\left.\begin{array}{ll} \text{if } \Delta \nu < 0 & \text{when } p \nearrow, p^{\Delta \nu} \searrow, \text{ equilibrium } \rightarrow \\ \text{if } \Delta \nu > 0 & \text{when } p \nearrow, p^{\Delta \nu} \nearrow, \text{ equilibrium } \leftarrow \\ \text{if } \Delta \nu = 0 & \text{equilibrium is not affected} \end{array}\right\} \tag{10.47}$$

10 Chemical Equilibria

The pressure term is affected by the stoichiometry of the reaction. If the pressure is increased, the system attempts to reduce the number of moles it contains, thus trying to oppose the change.

Example

Consider the reaction:

$$N_2 + 3H_2 \rightleftharpoons 2NH_3 \qquad \Delta \nu = \sum_i \nu_i = -2$$

The equilibrium is shifted to the right by a pressure increase. At 498.15 K, the standard equilibrium constant is $K^{\ominus}_{498.15} = 0.105$. In a system where we introduce 2 moles of H_2 and 1 mole of N_2, we find the following equilibrium conditions, which illustrate the equilibrium shift:

p_{total} [bar]	p_{N_2} [bar]	p_{H_2} [bar]	p_{NH_3} [bar]
1.00	0.319	0.596	0.084
10.00	2.72	3.61	3.67

10.8.3 Effect of Volume

To obtain the effect of a change in the volume of the system, we use the law of mass action expressed with mole numbers (according to Eq. 10.17).

$$K^{\ominus} = \left(\frac{RT}{p^{\ominus} V}\right)^{\Delta \nu} \prod_i n_i^{\nu_i} \qquad (10.48)$$

Again, the stoichiometry of the system controls the observed effects. From the above expression, we find that the following possibilities occur:

$$\left. \begin{array}{l} \text{if } \Delta \nu < 0 \quad \text{when } V \nearrow, \left(\dfrac{RT}{p^{\ominus} V}\right)^{\Delta \nu} \nearrow, \text{ equilibrium } \leftarrow \\[1em] \text{if } \Delta \nu > 0 \quad \text{when } V \nearrow, \left(\dfrac{RT}{p^{\ominus} V}\right)^{\Delta \nu} \searrow, \text{ equilibrium } \rightarrow \\[1em] \text{if } \Delta \nu = 0 \quad \text{equilibrium is not affected} \end{array} \right\} \qquad (10.49)$$

The system tries to increase the number of moles present, thus opposing the pressure decrease caused by a volume increase.

10.8.4 Effect of the Addition of an Inert Gas

An inert gas does not participate in the reaction, but it is of interest to see what happens when an inert gas is added to a system. If the addition is performed at constant volume and temperature, the partial pressures of the species participating in the reaction are not modified. Equation 10.17 shows that there is absolutely no effect. If the addition is at *constant pressure and temperature*, then the system volume increases and we obtain conclusions similar to those of Eq. 10.49 :

$$\left.\begin{array}{l} \text{if } \Delta v < 0 \quad \text{when adding an inert gas, equilibrium } \leftarrow \\ \text{if } \Delta v > 0 \quad \text{when adding an inert gas, equilibrium } \rightarrow \\ \text{if } \Delta v = 0 \quad \text{equilibrium is not affected} \end{array}\right\} \quad (10.50)$$

Example

Consider the reaction :

$$C \text{ (graphite)} + CO_2 \rightleftharpoons 2CO \qquad \Delta v = \sum_j v_j = 2 - 1 = 1$$

Only the gas phase species are written in the law of mass action. Keeping the system pressure constant and adding a gas that does not participate in the reaction, like nitrogen N_2, we find from Eq. 10.50 that the equilibrium is shifted to the right.

11. Perfect and Ideal Solutions

11.1 Basic Considerations

Consider a system containing a mixture of two species A and B that do not chemically react. Such a mixture, which contains only two species, is a **binary mixture**. The system, at equilibrium under the chosen experimental conditions, has *a single liquid phase and a gaseous phase*. The variance of this system is :

$$v = n + 2 - \varphi = 2 + 2 - 2 = 2 \tag{11.1}$$

For given values of the *temperature* of the system and of the *mole fraction* of one of the species in the liquid phase, all of the other intensive variables of the system are determined from equilibrium relations. The chemical potential of A in such a system is $\mu(p, T, x_A)$ where x_A is the mole fraction of A in the liquid. Using some very simple assumptions, one can find an explicit expression for the chemical potential of A.

At equilibrium, the partial pressure of a system containing only species A in the liquid state is $p_A^*(T,p)$, the vapor pressure of pure A which depends only slightly on the total pressure of the system (§ 8.5.4). If the system contained only pure B, then the partial pressure of A would be zero. A simple way to accommodate these two extreme cases is to assume that the partial pressure of A is proportional to x_A, the mole fraction of A in the liquid. With this assumption, we have :

$$p_A = x_A\, p_A^*(T, p) \simeq x_A\, p_A^*(T) \tag{11.2}$$

At equilibrium, the chemical potential of A has the same value in both phases. Assuming the gas phase behaves as an ideal gas, we have :

$$\left.\begin{aligned}
\mu_A(l) = \mu_A(g) &= \mu_A^\ominus(T) + R T \ln p_A \\
&= \mu_A^\ominus(T) + R T \ln p_A^*(T,p) + R T \ln x_A \\
\text{writing}\quad \mu_A^*(T,p) &= \mu_A^\ominus(T) + R T \ln p_A^*(T,p) \\
&\Downarrow \\
\mu_A(l) \quad &= \mu_A^*(T,p) + R T \ln x_A
\end{aligned}\right\} \tag{11.3}$$

$\mu_A^*(T, p)$ is the chemical potential of pure liquid A at T and p, which is simply related to the standard chemical potential of the vapor pressure of pure A. A species i in a solution is said to be *ideal* when its chemical potential is given by:

$$\mu_i = \mu_i^*(T, p) + RT \ln x_i \qquad (11.4)$$

where x_i is the mole fraction of species i in the solution. The standard chemical potential $\mu_i^*(T, p)$ corresponds to the value of the chemical potential for a mole fraction $x_i = 1$. The superscript $*$ indicates that the value is for the pure substance.

A *solution is ideal* if all of its species are ideal in some range of compositions.

Assuming the gas phase behaves as an ideal gas mixture, we obtain interesting results. At equilibrium, the chemical potential of each species is the same in the gas phase and in the liquid phase. We get:

$$\mu_A(l) = \mu_A(g) \Rightarrow \mu_A^*(T, p) + RT \ln x_A = \mu_A^\ominus(T) + RT \ln p_A \qquad (11.5)$$

from this equation, we can write:

$$\left.\begin{array}{l} p_A = x_A K_A \\ \text{with} \\ K_A = \exp\left(\dfrac{\mu_A^*(T, p) - \mu_A^\ominus(T)}{RT}\right) \end{array}\right\} \qquad (11.6)$$

We get back more than assumed, since we obtain the dependence of K_A on p. We have (6.25):

$$\left(\frac{\partial \ln K_A}{\partial p}\right)_T = \frac{1}{RT}\left(\frac{\partial \mu_A^*(T, p)}{\partial p}\right)_T = \frac{\overline{V_A}(l)}{RT} = \frac{V_{A,m}^*}{RT} \qquad (11.7)$$

where $V_{A,m}^*$ is the molar volume of pure A liquid. The variation of K_A with pressure is rather small since, in general, the molar volume of a liquid is small compared to RT.

Example

For H_2O at $25°C$, we have:

$$V_m = 18 \cdot 10^{-6} \, m^3 \qquad \left(\frac{\partial \ln K_A}{\partial p}\right)_T = 7.3 \cdot 10^{-9} \, Pa^{-1}$$

For a pressure change of 1 bar (10^5 Pa), the relative change of K_A is $7.3 \cdot 10^{-4}$.

When x_A is close to 1, the solution is ideal and we see that K_A corresponds to the *vapor pressure* of liquid A pure at pressure p. We can write the following relations neglecting the effect of the pressure on the chemical potential of pure A:

$$\left.\begin{array}{l} \mu_A^*(T,p) \simeq \mu_A^\ominus(T) + RT \ln p_A^*(T) \\ K_A \simeq p_A^*(T) = k_{RA} \quad \Leftarrow \text{ Raoult's constant} \\ \Downarrow \\ p_A \simeq x_A p_A^* \quad \Leftarrow \text{ Raoult's Law} \end{array}\right\} \qquad (11.8)$$

A species in a solution is said to obey **Raoult's law** when the last equation of 11.8 is experimentally verified. This type of behavior is experimentally observed in the case of liquid mixtures of species of a similar nature as well as for the solvent in dilute solutions. The constant k_{RA} is **Raoult's constant** for species A. It is equal to the vapor pressure of pure A. The standard chemical potential of A in the liquid is related to the standard chemical potential of A in the gas phase and to the vapor pressure of A by the first equation of 11.8.

Example

At 320 K, the vapor pressure of pure liquid hexane in equilibrium with its vapor is 48248 Pa. In an ideal solution with a mole fraction of hexane $x_{hex} = 0.4$, the partial pressure of hexane when liquid vapor equilibrium is achieved is :

$p_{hex} = 0.4 \cdot 48248 = 19299$ Pa

11.2 Perfect Solution

11.2.1 Isothermal Representation

When expression 11.4 for the chemical potential is valid for *all the species in the solution whatever the composition*, the solution is known as a **perfect solution**. Raoult's law is then valid for all species assuming the vapor behaves as an ideal gas mixture. We obtain the pressure of a binary system *at equilibrium* at a given temperature as a function of x_A :

$$\left.\begin{array}{l} p_A = x_A p_A^* \qquad p_B = x_B p_B^* \\ p = p_A + p_B = x_A p_A^* + x_B p_B^* \\ \text{with} \quad x_A + x_B = 1 \end{array}\right\} \Rightarrow p = p_B^* + x_A(p_A^* - p_B^*) \quad (11.9)$$

On Fig. 11.1, we display the data for a system where A is hexane and B is triethylamine at 350 K. The total equilibrium pressure is represented as a function of the composition of the liquid, known as the **vaporization curve** or **boiling point** (also **bubble point**) curve. We also display the partial pressures of the species as a function of the liquid composition. The *compositions of the vapor phase*, in A and B, y_A and y_B are :

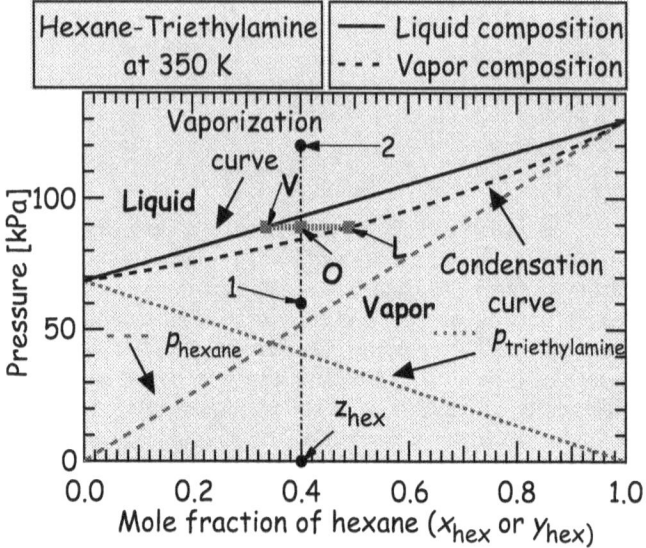

Figure 11.1 Partial vapor pressures and total pressure in a binary system at 350 K. The system contains hexane and triethylamine.

$$y_A = \frac{p_A}{p} = \frac{x_A p_A^*}{x_A p_A^* + x_B p_B^*} \qquad y_B = \frac{p_B}{p} = \frac{x_B p_B^*}{x_A p_A^* + x_B p_B^*} \qquad (11.10)$$

The curve giving the pressure of the system at equilibrium as a function of the vapor composition is the **condensation curve** or the **dew point curve**.

Example

In the data presented in Fig. 11.1, the vapor pressures of the pure species (hexane and triethylamine) at 350 K are $p_A^* = 129761$ Pa and $p_B^* = 68429$ Pa. For a composition of the liquid phase where $x_{hex} = 0.6$, the equilibrium pressure is :

$p = 0.6 \cdot 129761 + 0.4 \cdot 68429 = 105229$ bar

The vapor composition is :

$$y_{hex} = \frac{0.6 \cdot 129761}{105229} = 0.740$$

We have :

$$\frac{y_A}{y_B} = \frac{p_A}{p_B} = \frac{x_A p_A^*}{x_B p_B^*} \Rightarrow \begin{cases} \dfrac{y_A}{y_B} < \dfrac{x_A}{x_B} & \text{if } p_A^* < p_B^* \\ \dfrac{y_A}{y_B} > \dfrac{x_A}{x_B} & \text{if } p_A^* > p_B^* \end{cases} \qquad (11.11)$$

11 Perfect and Ideal Solutions

We see that the vapor is richer in the most volatile species.
For a given average composition of the system, z_A (here z_{hex}), we find that:

- If the point representing the state of the system (p and composition z_A) is below the condensation curve, the system only has a vapor (g) phase of composition z_A (point **1**).

- If the point representing the state of the system (p and composition z_A) is above the vaporization curve, the system has only a liquid phase of composition z_A (point **2**).

- If the point representing the system is between the two curves, then the system, at equilibrium, has one liquid and one gas phase (point **O**).

The ratio of the number of moles in the vapor (g) phase and the liquid (l) phase can be determined using the following relations:

$$\left. \begin{array}{r} n\, z_A = (n(l) + n(g))\, z_A \\ = n(l)\, x_A + n(g)\, y_A \end{array} \right\} \Rightarrow n(l)\,(z_A - x_A) = n(g)\,(y_A - z_A) \\ \Downarrow \\ \frac{n(g)}{n(l)} = \frac{(z_A - x_A)}{(y_A - z_A)} = \frac{\overline{OL}}{\overline{OV}} \qquad (11.12)$$

This rule is known as the **lever rule** and the result can be roughly read off the graph.

Example

A system containing 0.04 mol of hexane and 0.06 mol of triethylamine is at equilibrium at 89000 Pa and 350 K. At that pressure we have:

$$p = x_{hex} \cdot 129761 + (1 - x_{hex}) \cdot 68429 = 89000 \quad \Rightarrow \quad x_{hex} = 0.335$$

The composition of the vapor phase is:

$$y_{hex} = \frac{x_{hex}\, p^*_{hex}}{p} = \frac{0.335 \cdot 129761}{89000} = 0.489$$

We find the ratio of the number of moles in the vapor phase to the number of moles in the liquid phase to be:

$$\frac{n(g)}{n(l)} = \frac{(z_{hex} - x_{hex})}{(y_{hex} - z_{hex})} = \frac{0.4 - 0.335}{0.489 - 0.4} \simeq 0.73$$

All these results can be graphically estimated on Fig. 11.1.

11.2.2 Isobaric Representation

We can similarly select a constant system pressure, p, and the composition of the liquid phase. The other intensive variables are then determined.

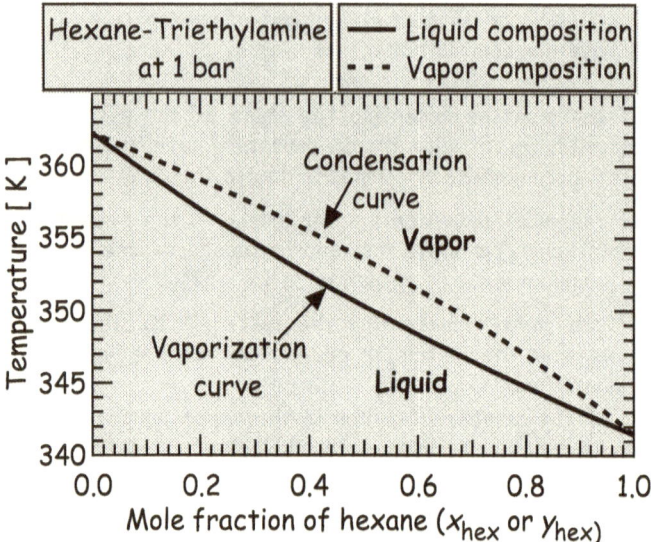

Figure 11.2 Equilibrium temperature as a function of the composition. x_{hex} is the mole fraction of hexane in the liquid, y_{hex} the mole fraction of hexane in the vapor. The total pressure is 1 bar.

The *vaporization* (or *boiling point* or *bubble point*) *curve* indicates the temperature at which the system pressure reaches p as a function of the mole fraction of hexane in the liquid.

The *condensation* (or *dew point*) *curve* gives the corresponding composition of the vapor. These curves are represented in Fig. 11.2, for a system containing the same species as that of Fig. 11.1. The vapor pressure is 1 bar at 341.8 K for pure hexane and at 362.7 K for pure triethylamine, a higher temperature since this species is less volatile.

The domain of stability of the liquid phase only, below the vaporization curve, corresponds to low temperatures. The domain of stability of the vapor phase only, above the condensation curve, corresponds to high temperatures. Equilibrium between a gas phase and a liquid phase is possible when the point representing the average composition of the system is between both curves. The vaporization curve is no longer a straight line. Its shape is related to the change of the vapor pressure of the species with temperature. A sequence of distillations followed by condensations leads to fractions of various composition, providing a simple method of purification of liquid products, for which Raoult's law is valid. This is not necessarily possible for solutions that are not perfect (see chapter 12). The ratio of the number of moles in the gas phase to the number of moles in the liquid phase can also be determined using the *lever rule*.

Example

At 345 K, the vapor pressures of pure hexane and pure triethylamine are p^*_{hex} = 111575 Pa and p^*_{trieth} = 58070 Pa. For a total pressure of 1 bar, we obtain the compositions of the liquid and vapor phases. We have :

$$p_{total} = x_{hex} p^*_{hex} + (1 - x_{hex}) p^*_{trieth} = 1 \text{ bar} = 10^5 \text{ Pa}$$

$$x_{hex} = \frac{p^*_{trieth} - p_{total}}{p^*_{trieth} - p^*_{hex}} = \frac{58070 - 10^5}{58070 - 111575} \simeq 0.784$$

The vapor composition is :

$$y_{hex} = \frac{x_{hex} p^*_{hex}}{p_{total}} = 0.874$$

11.3 Mixing Properties of Ideal Solutions

Consider a number of small systems, each made up of one pure compound, liquid at temperature T and pressure p. The partial molar volume of each pure species is equal to its molar volume, so that the system made of this ensemble of pure compounds has a volume :

$$V_{initial} = \sum_i n_i V_{i,m} \qquad (11.13)$$

Assume the species do not react when they are mixed. We mix them at constant temperature and pressure and assume that the resulting solution is *ideal*. The chemical potential of each of the species in the system is given by an equation such as 11.4. Using Eq. 6.25, the partial molar volume of each of the species in the solution is given by :

$$\overline{V_i} = \left(\frac{\partial \mu_i(l)}{\partial p}\right)_{T, n_i, n_j} = \left(\frac{\partial \mu^*_i}{\partial p}\right)_T = V^*_{i,m}(l) = V_{i,m} \qquad (11.14)^\dagger$$

since μ^*_i is independent of composition. This result leads to :

$$V_{final} = \sum_i n_i \overline{V_i} = \sum_i n_i V_{i,m} = V_{initial} \Rightarrow \Delta_{mix} V = 0 \qquad (11.15)$$

The volume of mixing of an ideal solution is zero. A similar result is obtained for enthalpy using 6.29 :

$$\overline{H_i} = -T^2 \left[\frac{\partial}{\partial T}\left(\frac{\mu_i}{T}\right)\right]_{p, n_i, n_j} = -T^2 \left[\frac{\partial}{\partial T}\left(\frac{\mu^*_i}{T}\right)\right]_p = H^*_{i,m}(l) = H_{i,m}$$

$$(11.16)$$

---

†$V^*_{i,m}(l)$ means that we are dealing with the molar volume of species i as a pure liquid. It is therefore exactly equal to $V_{i,m}$.

which implies that:

$$H_{final} = \sum_i n_i \overline{H_i} = \sum_i n_i H_{i,m} = H_{initial} \Rightarrow \Delta_{mix}H = 0 \quad (11.17)$$

These results imply that the internal energy of mixing, $\Delta_{mix}U$, is also zero. Some of the mixing variables are not zero. First, we express the partial molar entropy of species i:

$$\overline{S_i} = -\left(\frac{\partial \mu_i(l)}{\partial T}\right)_{p, n_i, n_j} = -\left(\frac{\partial \mu_i^*}{\partial T}\right)_p - R \ln x_i = S_{i,m}^*(l) - R \ln x_i \quad (11.18)$$

The entropy and the Gibbs energy of mixing are:

$$\left.\begin{array}{l}\Delta_{mix}S = \sum_i n_i (S_{i,m}^*(l) - R \ln x_i) - \sum_i n_i S_{i,m}^*(l) \\ = -R \sum_i n_i \ln x_i\end{array}\right\} \quad (11.19)$$

$$\Delta_{mix}G = \sum_i n_i \mu_i - \sum_i n_i \mu_i^* = RT \sum_i n_i \ln x_i \quad (11.20)$$

Example

At 300 K and a pressure of 1 bar, we mix 0.1 mol of hexane with 0.2 mol of triethylamine. At that pressure, the system is purely liquid. The entropy of mixing is:

$$\Delta_{mix}S = -R \sum_i n_i \ln x_i = -8.3145 \left(0.1 \ln \frac{0.1}{0.3} + 0.2 \ln \frac{0.2}{0.3}\right) \simeq 1.588 \text{ J K}^{-1}$$

and the Gibbs energy of mixing is:

$$\Delta_{mix}G = -T\Delta_{mix}S \simeq -476 \text{ J}$$

11.4 Effect of Pressure and Temperature on Liquid Vapor Equilibria

We consider an ideal solution in equilibrium with its vapor (assumed to be an ideal gas mixture). The chemical potential of one of its species, i, has the same value in the gas phase and the liquid phase. We have:

$$\left.\begin{array}{l}\mu_i = \mu_i^*(T,p) + RT \ln x_i = \mu_i^\ominus(T) + RT \ln p_i \\ \Downarrow \\ \ln \frac{p_i}{x_i} = \left(\frac{\mu_i^*(T,p) - \mu_i^\ominus(T)}{RT}\right)\end{array}\right\} \quad (11.21)$$

11 Perfect and Ideal Solutions

The right hand side of Eq. 11.21 depends only on p and T. Taking into account the fact that the standard chemical potential of the gas is independent of pressure, we obtain:

$$d\left(\ln \frac{p_i}{x_i}\right) = \frac{1}{R}\left[\frac{\partial\left(\frac{\mu_i^* - \mu_i^\ominus}{T}\right)}{\partial T} dT + \frac{\partial\left(\frac{\mu_i^*}{T}\right)}{\partial p} dp\right]$$

$$= \frac{H_i^\ominus(g) - H_{i,m}^*(l)}{RT^2} dT + \frac{V_{i,m}^*(l)}{RT} dp$$

(11.22)

We saw in § 11.1 that the numerical value of the coefficient of dp is often quite small. The ratio of the partial pressure of one of the species in solution to its mole fraction is barely affected by a pressure change (as already mentioned). The variation of the partial pressure of a species with temperature depends mainly on its latent heat of vaporization.

Example

At 300 K, the latent heat of vaporization of water is about 44000 J. We find:

$$\frac{H_i^\ominus(g) - H_{i,m}^*(l)}{RT^2} = \frac{44000}{8.3145 \cdot (300)^2} \simeq 0.06 \quad \Rightarrow \quad \exp(0.06) \simeq 1.06$$

At constant pressure, a 1 K change corresponds to a relative change of the ratio p/x of about 6%.

11.5 Lowering of the Freezing Temperature of a Solvent in the Presence of a Solute – Eutectic

When in a liquid solution, one of the species has a high mole fraction, it is designated as the **solvent**. The other species present at smaller mole fractions are known as the **solutes**. We assume that we know the freezing temperature T_0 of a pure solvent at pressure p. Let us examine how the solid-liquid equilibrium is affected by the presence of a solute, when the solvent behaves ideally (this is always the case for dilute solutions §12.6). When the solution contains both *pure solid* and *liquid solvent* and is at equilibrium, the chemical potentials of the liquid and solid forms of the solvent have the same value:

$$\left.\begin{array}{l}\mu_{solv}(l) = \mu_{solv}(s) \Rightarrow \mu_{solv}^*(T,p) + RT \ln x_{solv} = \mu_{solv}^\ominus(s) \\ \ln x_{solv} = \frac{1}{RT}[\mu_{solv}^\ominus(s) - \mu_{solv}^*(T,p)]\end{array}\right\}$$

(11.23)

We assume here that the chemical potential of the solid is independent of pressure and equal to its standard potential and we find the change in x_{solv} with temperature:

$$\left(\frac{\partial \ln x_{solv}}{\partial T}\right)_p = \frac{1}{R}\left[\frac{\partial \left(\frac{\mu^{\ominus}_{solv}(s)}{T}\right)}{\partial T} - \frac{\partial \left(\frac{\mu^{*}_{solv}(T,p)}{T}\right)}{\partial T}\right]$$

$$= \frac{1}{R}\left[-\frac{H_{solv}(s)}{T^2} + \frac{H^{*}_{m,solv}(l)}{T^2}\right] \qquad (11.24)$$

$$= \frac{H^{*}_{m,solv}(l) - H^{*}_{m,solv}(s)}{RT^2} = \frac{L_{solv}(T)}{RT^2}$$

where L_{solv} is the *latent heat of fusion* of the solvent at temperature T. By integrating Eq. 11.24, we can obtain the change in the temperature at which the solvent freezes when a solute is added to it. We get:

$$\int_1^{x_{solv}} d\ln x_{solv} = \int_{T_0}^{T} \frac{L_{solv}(T)}{RT^2} dT \Rightarrow \ln x_{solv} \simeq \frac{L_{solv}(T_0)}{R}\left(\frac{1}{T_0} - \frac{1}{T}\right) \quad (11.25)$$

The right hand side of the second Eq. 11.25 is obtained by assuming the latent heat of fusion of the solvent is independent of temperature, which is a valid assumption when the freezing temperature changes observed are fairly small. When the mole fraction of the solvent changes in the liquid phase by addition of a solute, the freezing temperature of the solvent changes. It is remarkable that this variation is independent of the characteristics of the solute. Moreover, if the mole fraction of solute present is small as well, the solvent mole fraction is close to 1 and we can write the following approximations:

$$x_{solv} = 1 - x_{solute} \quad \text{and} \quad \ln(1 - x_{solute}) \simeq -x_{solute}$$
$$\Downarrow$$
$$x_{solute} \simeq \frac{L_{solv}(T_0)}{R} \frac{T_0 - T}{TT_0} \qquad (11.26)$$

Since the latent heat of fusion $L_{solv}(T_0)$ of the solvent is always a positive quantity, we always have $T_0 > T$, *the freezing temperature is always lowered by the presence of a solute*. By measuring the lowering of the freezing temperature, it is therefore possible to obtain the mole fraction of the solute present in the solution. If the mass of the amount introduced is known, it is possible to obtain the molar mass of the substance.

11 Perfect and Ideal Solutions

Example

Let us find the lowering of the freezing point of an aqueous solution containing 68% of ethanol by weight. The latent heat of fusion of water is 6008 J. The molecular weight of the two species are $M_{H_2O} = 18.015$ and $M_{EtOH} = 46.07$. In 100 g of mixture, we have:

$$\left. \begin{array}{l} \dfrac{(100-68)}{18.015} = 1.78 \text{ mol of water} \\ \dfrac{68}{46.07} = 1.48 \text{ mol of ethanol} \end{array} \right\} \Rightarrow x_{solute} = \dfrac{1.48}{1.48+1.78} \simeq 0.454$$

$$\ln x_{solv} = \ln(1-0.454) \simeq \dfrac{L_{solv}(T_0)}{R}\left(\dfrac{1}{T_0} - \dfrac{1}{T}\right) = \dfrac{6008}{8.3145}\left(\dfrac{1}{273.15} - \dfrac{1}{T}\right)$$

Solving this equation, we find that freezing of the water should take place at 222.3 K, which corresponds to a lowering of the freezing temperature of 50.8 K. This estimate is excellent since the observed lowering is 49.5K.

Consider a binary system at a *given pressure* where the solid phases are pure species and solubility of both species is complete over the entire range of composition. When the temperature is lowered and one of the species freezes out, the composition of the solution varies. The variance of such a system is $v = 2+2-2 = 2$ (p and T). Equation 11.25 provides the equilibrium curves on a T vs. x graph, considering either of the species as the solvent. At the point where these two curves meet, the system at equilibrium has two solid and one liquid phases. The variance is $v = 2+2-3 = 1$. For a given p, this point is unique and is called the **Eutectic point**. It is the lowest temperature where a solid liquid equilibrium is possible at p.

Example

	T_{fusion} [K]	Enthalpy of fusion [J mol^{-1}]
Benzene	278.68	9950
Naphthalene	353.35	19060

We consider mixtures of benzene and naphthalene. Using the above data, we obtain the equilibrium curves:

$$T = \dfrac{L_{fus\ Napht}\ T_{0\ Napht}}{L_{fus\ Napht} - T_{0\ Napht}\ R \ln x_{Napht}}$$

$$T = \dfrac{L_{fus\ Benz}\ T_{0\ Benz}}{L_{fus\ Benz} - T_{0\ Benz}\ R \ln x_{Benz}}$$

For temperatures below the Eutectic line, the system is solid. Above the equilibrium curves, the system is liquid. We indicate the phases present in the different zones delimited by the lines. Eutectic: $x_{Napht} \simeq 0.133$ at $T \simeq 269.8$ K.

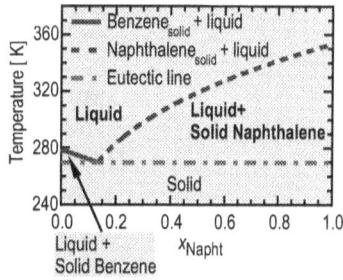

11.6 Elevation of the Boiling Temperature of a Solvent in the Presence of a Non Volatile Solute

We now consider the case of a solvent that boils at temperature T_0 when it is pure and at pressure p. It is equivalent to say that, at temperature T_0, the vapor pressure of the pure solvent is p. Let us investigate how a *non volatile* solute affects the vapor liquid equilibrium when the solvent behaves ideally. The analysis is very similar to that of § 11.5. We write that the chemical potential of the solvent is the same in the gas phase and the liquid phase.

$$\left.\begin{array}{l} \mu_{solv}(l) = \mu_{solv}(g) \;\Rightarrow\; \mu^*_{solv}(T,p) + RT \ln x_{solv} = \mu_{solv}(g) \\[6pt] \left(\dfrac{\partial \ln x_{solv}}{\partial T}\right)_p = \dfrac{H^*_{m,\,solv}(l) - \overline{H}_{solv}(g)}{RT^2} = -\dfrac{L_{solv}(T)}{RT^2} \\[6pt] \overline{H}_{solv}(g) = H^\ominus_{solv}(g) \qquad \text{if vapor is an ideal gas} \end{array}\right\}$$

(11.27)

where $L_{solv}(T)$ represents, this time, the *latent heat of vaporization of the solvent* at pressure p and temperature T. The change in the boiling temperature is obtained by integration of Eq. 11.27 and the results are very similar to those obtained for the lowering of the freezing point.

$$\int_1^{x_{solv}} d \ln x_{solv} = \int_{T_0}^{T} -\dfrac{L_{solv}(T)}{RT^2} dT \;\Rightarrow\; \ln x_{solv} \simeq \dfrac{L_{solv}}{R}\left(\dfrac{1}{T} - \dfrac{1}{T_0}\right)$$

(11.28)

The right hand side of the second Eq. 11.28 is obtained by assuming the latent heat of vaporization of the solvent is independent of temperature, which is a valid assumption when the boiling temperature changes observed are fairly small. With the same type of approximations used for Eq. 11.26, we get:

$$\left.\begin{array}{c} x_{solv} = 1 - x_{solute} \quad \text{and} \quad \ln(1 - x_{solute}) \simeq -x_{solute} \\ \Downarrow \\ x_{solute} \simeq \dfrac{L_{solv}}{R}\dfrac{T - T_0}{T\,T_0} \end{array}\right\}$$

(11.29)

The latent heat of vaporization is always positive, thus we will always find that $T > T_0$, the boiling temperature is always raised in the presence of a non volatile solute. If a known mass of a single solute is

11 Perfect and Ideal Solutions

present, one may obtain the molar mass of the substance from the elevation of the boiling temperature.

Example

Let us find the expected elevation of the boiling temperature of water when a 0.05 M solution of a non volatile solute is added. The enthalpy of vaporization of water at 373.15 K is 40657 J mol^{-1}. We have:

$$x_{solute} \simeq \frac{L_{solv}}{R} \frac{T-T_0}{T T_0} \quad \Rightarrow \quad T - T_0 = 0.05 \cdot (373.15)^2 \frac{8.3145}{40657} \simeq 1.424 \text{ K}$$

11.7 Osmotic Pressure

We now direct our interest to a system with two liquid phases. One is a *pure solvent*, while the other one is the *same solvent containing a solute*. The two phases are separated by a membrane permeable only to solvent molecules. Such a membrane is known as a semipermeable membrane.

This system is schematically represented in Fig. 11.3 and the chemical potential of the solvent is given by Eq. 11.4, for each part of the system. If both liquid phases have the same pressure, p, the chemical potential of the pure solvent is larger than that of the solvent containing the solute, since the mole fraction of the solvent is less than 1 in that case. The solvent will therefore tend to spontaneously pass through the membrane going from the high chemical potential region (pure solvent) to the lower chemical potential region (solution). The *osmotic pressure* is the pressure difference that must exist between the two compartments when they are at equilibrium. The flux of solvent is then zero. Let us assume that equilibrium is achieved when the pressure on the pure solvent side is p and the pressure on the solution side is p'.

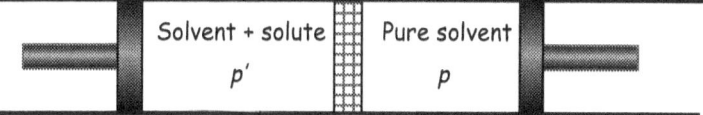

Figure 11.3 Osmotic pressure. Only pure solvent can go through the membrane separating the two compartments. At equilibrium, a pressure difference must exist between the two compartments.

The chemical potential of the solvent is then the same in both compartments, which we write:

$$\mu^*_{solv}(T, p') + R T \ln x_{solv} = \mu^*_{solv}(T, p) \qquad (11.30)$$

This expression can be transformed by making use of the fact that the molar volume of a liquid does not vary much with pressure.

$$-RT \ln x_{solv} = \mu^*_{solv}(T, p') - \mu^*_{solv}(T, p)$$

$$= \int_p^{p'} \left(\frac{\partial \mu^*_{solv}}{\partial p}\right)_T dp = \int_p^{p'} V^*_{m, solv} \, dp \qquad (11.31)$$

$$\simeq V^*_{m, solv}(p' - p) = V^*_{m, solv} \, \Pi$$

The excess pressure Π, the osmotic pressure, needs to be applied to the solute containing side to achieve equilibrium between the two compartments. It depends on the mole fraction of the solvent in the compartment that contains the solute. If only one solute is present and its mole fraction is small, then we can write the following approximations:

$$x_{solv} = 1 - x_{solute} \quad \text{and} \quad \ln(1 - x_{solute}) \simeq -x_{solute} \simeq -\frac{n_{solute}}{n_{solv}}$$

⇓

$$\Pi = p' - p \simeq \frac{RT}{V^*_{m, solv}} x_{solute} \simeq \frac{n_{solute} \, RT}{V_{solv}}$$

(11.32)

This equation bears a certain resemblance to the equation of state of an ideal gas.

Example

Let us evaluate the osmotic pressure at 300 K for a system where we have a solution of 10 g of methanol in 100 g of water separated from pure water by a membrane permeable only to water. We use $M_{H_2O} = 18.05$, $M_{MeOH} = 32.04$ and $V_m(H_2O) = 18.11 \, 10^{-6} \, m^3 \, mol^{-1}$. We have:

$$\left. \begin{array}{l} n_{H_2O} = \dfrac{100}{18.05} \simeq 5.54 \text{ mol} \\ n_{MeOH} = \dfrac{10}{32.04} \simeq 0.312 \text{ mol} \end{array} \right\} \Rightarrow x_{solv} = \dfrac{n_{H_2O}}{n_{H_2O} + n_{MeOH}} = 0.947$$

We find that the osmotic pressure is:

$$\Pi = \frac{-8.3145 \cdot 300 \ln(0.947)}{18.11 \, 10^{-6}} \simeq 7.5 \, 10^6 \text{ Pa} = 75 \text{ bar}$$

12. Non Ideal Solutions

12.1 Introduction

Most real solutions behave neither perfectly nor ideally. It is convenient to keep, as an expression for the chemical potential of species in non ideal solutions, an expression formally similar to that used for ideal or perfect solutions. The chemical potential of species i in a non ideal solution at temperature T and pressure p is expressed as:

$$\mu_i(l) = \mu^*_{x\,i}(T, p) + R\,T \ln a_{x\,i} = \mu^*_{x\,i}(T, p) + R\,T \ln \gamma_{x\,i}\, x_i \quad (12.1)$$

where $a_{x\,i}$ is known as the **activity** of species i. The coefficient $\gamma_{x\,i}$ is the **activity coefficient** of species i in the solution. The subscript x indicates that the composition scale selected is mole fractions. We now suppress this subscript to simplify the notation, remembering that when no subscript is mentioned, then it is agreed that the composition scale used is mole fractions. We write:

$$\mu_i(l) = \mu^*_i(T, p) + R\,T \ln \gamma_i\, x_i \quad (12.1')$$

The coefficient γ_i describes the deviation of the chemical potential of species i in the real solution from what it would be in an ideal solution when present at the same mole fraction. This activity coefficient depends on pressure, temperature and composition of the solution $x_1,..., x_i,...$. The chemical potential has the value $\mu^*_i(T, p)$ when the activity $a_{x\,i}$ has a value of 1. Later in this chapter, we will see that several conventions may be selected for the activity coefficient and consequently several standard states.

When a solution is in equilibrium with a gas phase, species i has the same chemical potential in both phases. Assuming that the gas phase behaves as an ideal gas, we have:

$$\mu_i(l) = \mu_i(g) \Rightarrow \mu^*_i(T, p) + R\,T \ln \gamma_i\, x_i = \mu^\ominus_i(T) + R\,T \ln p_i \quad (12.2)$$

which can be written as:

$$\left. \begin{array}{l} p_i = \gamma_i\, x_i\, k_i \\ \text{where} \\ k_i = \exp\left(\dfrac{\mu^*_i(T, p) - \mu^\ominus_i(T)}{R\,T} \right) \end{array} \right\} \quad (12.3)$$

The constant k_i is independent of the composition of the system. It depends on temperature and pressure (small dependence with respect to the latter) and on the convention selected for the activity coefficient and for the liquid reference state, which affects $\mu_i^*(T, p)$. A precise interpretation of the significance of Eq. 12.3 depends on the convention selected for the standard state (See § 12.4, 12.6).

Example
Consider a mixture of water and propanol at 40°C in equilibrium with its vapor. The equilibrium pressure is 10977 Pa, the composition of the liquid phase is $x_{prop} = 0.7$ while that of the vapor is $y_{prop} = 0.472$. As we will see in § 12.4, when we select convention I for the activity coefficients, the coefficient k_i for Eq. 12.3 is the vapor pressure of the species, $p^*_{prop} = 6774$ Pa and $p^*_{H_2O} = 7377$ Pa.
The activity coefficients (convention I) for both species are:

$$\gamma^I_{prop} = \frac{p_{prop}}{x_{prop} p^*_{prop}} = \frac{y_{prop} p_{eq}}{x_{prop} p^*_{prop}} = \frac{0.472 \cdot 10977}{0.7 \cdot 6774} = 1.093$$

$$\gamma^I_{H_2O} = \frac{p_{H_2O}}{x_{H_2O} p^*_{H_2O}} = \frac{y_{H_2O} p_{eq}}{x_{H_2O} p^*_{H_2O}} = \frac{0.528 \cdot 10977}{0.3 \cdot 7377} = 2.618$$

12.2 Variables and Excess Variables of Mixing

The **Gibbs energy of mixing** is given by:

$$\Delta_{mix} G = \sum_i n_i \mu_i - \sum_i n_i \mu_i^* = RT \sum_i n_i \ln \gamma_i x_i \qquad (12.4)$$

The difference between the Gibbs energy of mixing for a real solution and the Gibbs energy of mixing if the mixture was ideal is the **excess Gibbs energy of mixing**. It is given by:

$$G^E = \Delta_{mix} G - \Delta_{mix} G_{ideal} = RT \sum_i n_i \ln \gamma_i \qquad (12.5)$$

Example
For one mole of the mixture considered in the example of § 12.1, the Gibbs energy of mixing is:

$\Delta_{mix} G = RT(n_{prop} \ln \gamma_{prop} x_{prop} + n_{H_2O} \ln \gamma_{H_2O} x_{H_2O})$
$= 8.3145 \cdot 313.15 \cdot (0.7 \ln(1.093 \cdot 0.7) + 0.3 \ln(2.618 \cdot 0.3))$
$= -677$ J

If the system were ideal, the Gibbs energy of mixing would be:

$\Delta_{mix} G_{ideal} = RT(n_{prop} \ln x_{prop} + n_{H_2O} \ln x_{H_2O})$
$= 8.3145 \cdot 313.15 \cdot (0.7 \ln(0.7) + 0.3 \ln(0.3))$
$= -1591$ J

12 Non Ideal Solutions

The excess Gibbs energy of mixing is:

$$G^E = RT \sum_i n_i \ln \gamma_i = 8.3145 \cdot 313.15 \cdot (0.7 \ln(1.093) + 0.3 \ln(2.618)) = 914 \text{ J}$$

The expression for the *partial molar entropy* of species *i* is:

$$\begin{aligned}
\overline{S_i} &= -\left(\frac{\partial \mu_i(l)}{\partial T}\right)_{p, n_i, n_j} \\
&= -\left(\frac{\partial \mu_i^*}{\partial T}\right)_p - R \ln x_i - R \ln \gamma_i - RT\left(\frac{\partial \ln \gamma_i}{\partial T}\right)_{p, n_i, n_j} \\
&= S_{i,m}^*(l) - R \ln x_i - R \ln \gamma_i - RT\left(\frac{\partial \ln \gamma_i}{\partial T}\right)_{p, n_i, n_j}
\end{aligned} \quad (12.6)$$

Using Eq. 12.6, we can obtain the *entropy of mixing*:

$$\begin{aligned}
\Delta_{mix}S &= \sum_i n_i \overline{S_i} - \sum_i n_i S_{i,m}^*(l) \\
&= -\sum_i n_i \left(R \ln x_i + R \ln \gamma_i + RT\left(\frac{\partial \ln \gamma_i}{\partial T}\right)_{p, n_i, n_j}\right)
\end{aligned} \quad (12.7)$$

and the *excess entropy of mixing* is given by:

$$S^E = \Delta_{mix}S - \Delta_{mix}S_{ideal} = -\sum_i n_i \left(R \ln \gamma_i + RT\left(\frac{\partial \ln \gamma_i}{\partial T}\right)_{p, n_i, n_j}\right)$$

(12.8)

12.3 Effect of Pressure and Temperature on the Activity Coefficient

We can start from expression 12.1' for the chemical potential of a species in a system and, using Eq. 6.30, we obtain:

$$\begin{aligned}
\left(\frac{\partial \mu_i(l)}{\partial p}\right)_{T, n_i, n_j} &= \overline{V_i} = \left(\frac{\partial \mu_i^*(T, p)}{\partial p}\right)_{T, n_i, n_j} + RT\left(\frac{\partial \ln \gamma_i}{\partial p}\right)_{T, n_i, n_j} \\
&= V_{i,m}^* + RT\left(\frac{\partial \ln \gamma_i}{\partial p}\right)_{T, n_i, n_j}
\end{aligned}$$

(12.9)

$V_{i,m}^*$ is the molar volume of pure species *i* in the liquid state at temperature *T* and at pressure *p*. The variation of the activity coefficient with pressure is:

$$\left(\frac{\partial \ln \gamma_i}{\partial p}\right)_{T, n_i, n_j} = \frac{\overline{V_i} - V^*_{i,m}}{RT} \quad (12.10)$$

The difference $\overline{V_i} - V^*_{i,m}$ is often quite small compared to RT and any variation with pressure of the activity coefficients can often be neglected. For the effect of temperature:

$$\left[\frac{\partial}{\partial T}\left(\frac{\overline{\mu_i}}{T}\right)\right]_{p, n_i, n_j} = -\frac{\overline{H_i}}{T^2} = \left[\frac{\partial}{\partial T}\left(\frac{\mu^*_i}{T}\right)\right]_{p, n_i, n_j} + R\left(\frac{\partial \ln \gamma_i}{\partial T}\right)_{p, n_i, n_j}$$
$$= -\frac{H^*_{i,m}}{T^2} + R\left(\frac{\partial \ln \gamma_i}{\partial T}\right)_{p, n_i, n_j} \quad (12.11)$$

where $H^*_{i,m}$ (Eq. 6.29) is the molar enthalpy of pure species i in the liquid state at temperature T and at pressure p. We get:

$$\left(\frac{\partial \ln \gamma_i}{\partial T}\right)_{p, n_i, n_j} = -\frac{\overline{H_i} - H^*_{i,m}}{RT^2} \quad (12.12)$$

Measurement of partial molar volumes and partial molar enthalpies provides information on the variation of the activity coefficients with pressure and temperature. The result of Eq. 12.12 and 12.5 can be used in Eq. 12.8. to obtain the excess entropy of mixing:

$$-RT\sum_i n_i \left(\frac{\partial \ln \gamma_i}{\partial T}\right)_{p, n_i, n_j} = \sum_i \frac{\overline{H_i} - H^*_{i,m}}{T} = \frac{\Delta_{mix}H}{T}$$
$$S^E = \frac{\Delta_{mix}H - G^E}{T} \quad (12.13)$$

Example

For one mole of the mixture considered in the example of § 12.1, the enthalpy of mixing is measured to be 378 J. The excess entropy of mixing is therefore:

$$S^E = \frac{\Delta_{mix}H - G^E}{T} = \frac{378 - (914)}{313.15} \approx -1.71 \text{ J K}^{-1}$$

If the system were ideal, the entropy of mixing would have been:

$$\Delta_{mix}S_{ideal} = -R\sum_i n_i \ln x_i$$
$$= -8.3145 \cdot (0.7 \ln(0.7) + 0.3 \ln(0.3))$$
$$= 5.08 \text{ J K}^{-1}$$

12.4 Standard State – Convention I for the Activity Coefficient

Convention I for the activity coefficient and the standard state is particularly appropriate when all of the species are liquid in the conditions of pressure and temperature of the system. The activity coefficient γ_i^I is taken to have value 1 when the mole fraction of species i is 1. We have:

$$\left. \begin{array}{l} \mu_i = \mu_i^{*I}(T, p) + R\,T \ln \gamma_i^I x_i \\ \text{with } \gamma_i^I \to 1 \text{ when } x_i \to 1 \end{array} \right\} \quad (12.14)$$

The *chemical potential* $\mu_i^{*I}(T, p)$ corresponds to the chemical potential of species i pure and in the liquid state at T and p. In a dilute solution, the mole fraction of the solvent is close to 1, the solvent behaves ideally. When the pressure is the standard state pressure $p^\ominus$, $\mu_i^{*I}(T, p^\ominus)$ is the standard chemical potential of species i. When using convention I, Eq. 12.3 can be written:

$$\left. \begin{array}{l} p_i = \gamma_i^I x_i k_{Ri} \\ \text{where} \\ k_{Ri} = \exp\left(\dfrac{\mu_i^{*I}(T, p) - \mu_i^\ominus(T)}{R\,T} \right) \\ \dfrac{p_i}{x_i k_{Ri}} = \gamma_i^I \simeq \dfrac{p_i}{x_i p_i^*(T)} \end{array} \right\} \quad (12.15)$$

The coefficient k_{Ri} is **Raoult's constant**. It is equal to $p_i^*(T, p)$, the vapor pressure of species i at pressure p. As explicitly shown in Eq. 11.7, its dependence on pressure is small. Therefore Raoult's constant can be considered to be $p_i^*(T)$, the vapor pressure of the pure species. The pressure has also very little effect on related quantities that are given in tables for the standard state pressure $p^\ominus$. If the activity coefficient γ_i^I is larger than 1, the species is said to present a *positive deviation*. If the activity coefficient γ_i^I is smaller than 1, the species is said to present a *negative deviation*. For a perfect solution, the activity coefficient is 1 over the entire range of composition.

The **standard chemical potential** when using convention I is $\mu_i^{*I}(T, p^\ominus)$, the chemical potential of the species when it behaves ideally, is pure and at the standard state pressure, $p^\ominus$. In practice, this standard chemical potential is often noted as $\mu_i^\ominus(l)$.

The standard chemical potential of species *i* according to convention I is related to the standard chemical potential of gaseous *i*, assumed to be an ideal gas, $\mu_i^\ominus(T)$:

$$\mu_i^{*I}(T, p^\ominus) \simeq \mu_i^{*I}(T, p) = \mu_i^\ominus(T) + R\,T \ln p_i^*(T, p)$$

$$\simeq \mu_i^\ominus(T) + R\,T \ln p_i^*(T) \qquad (12.16)$$

Example

At 40°C, the vapor pressure of pure CCl_4 is 47730 Pa. The Raoult's constant of CCl_4 is:
$k_{R\,CCl_4} = 47730$ Pa

Raoult's constant is a characteristic of the pure species.

12.5 Liquid - Vapor Equilibrium

12.5.1 Isothermal Diagram

Consider a mixture of two miscible liquids in equilibrium with its vapor phase. Such a system is divariant ($v = 2 + 2 - 2$). For a system at a constant temperature *T*, there is an equilibrium pressure for each possible composition of the liquid mixture.

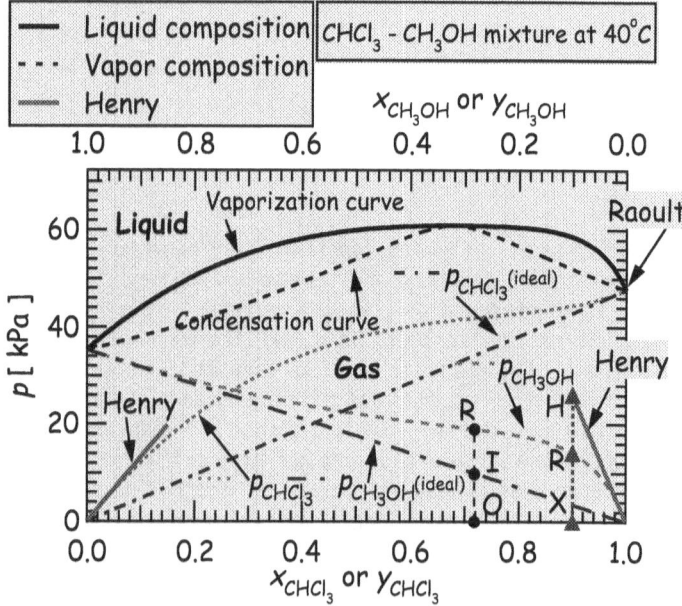

Figure 12.1 Isothermal diagram (at 40°C) of a non ideal liquid solution, a mixture of chloroform and methanol. A maximum azeotrope exists at $x_{CHCl_3} \simeq 0.68$ and $p \simeq 0.609$ bar.

12 Non Ideal Solutions

The corresponding composition of the vapor phase is usually represented on the same graph. Depending on the species present, the deviation with respect to the ideal case may be more or less pronounced. On Fig. 12.1, we have the equilibrium curves for a binary mixture of chloroform and methanol. It is very far from ideal.

For a representative point (pressure and composition) above the vaporization curve, the pressure in the system is sufficiently high for the entire system to be liquid. For a representative point located below the condensation curve, the entire system is gaseous. If the point representing the system is between the two curves, then the system, at equilibrium, has one liquid and one gas phase. The composition of the phases is determined by the intersection of a horizontal line going through the representative point and each of the curves. By extension, the lever rule discussed in chapter 11 also applies.

In the present case, the difference with the simple case of an ideal solution is so large that each curve presents an extremum (here a maximum). It can be shown that when this occurs, the compositions of the liquid and vapor phases are identical. This composition is called the **azeotropic composition**. The two species are said to form an **azeotrope**.

The partial pressure of each of the species as a function of composition is very different from what it should be if the solution were perfect.

For a liquid composition corresponding to point O, the activity coefficient of CH_3OH is given by the ratio of the measured partial pressure of CH_3OH to the corresponding ideal pressure. The activity coefficient is given by:

$$\gamma^I_{CH_3OH} = \frac{OR}{OI} > 1 \qquad (12.17)$$

The azeotrope presented here is a *maximum azeotrope* or *positive azeotrope*. (The pressure deviation is positive).

The activity coefficient of each of the two species A and B *for the azeotropic composition* is easily obtained by taking into account the fact that both phases have the same composition ($x_A = y_A$). Using convention I, the partial pressure of each of the species are:

$$\left. \begin{array}{l} p_A = y_A \, p = \gamma^I_A \, x_A \, p^*_A(T) \;\Rightarrow\; \gamma^I_A = \dfrac{p}{p^*_A(T)} \\[1em] p_B = y_B \, p = \gamma^I_B \, x_B \, p^*_B(T) \;\Rightarrow\; \gamma^I_B = \dfrac{p}{p^*_B(T)} \end{array} \right\} \qquad (12.18)$$

We find that, in this special case, the activity coefficient of a species is the ratio of the equilibrium pressure of the system to the vapor pressure of the pure species at the same temperature.

Example

For the chloroform methanol mixture at 40°C, the azeotrope composition is $x_{CHCl_3} \simeq 0.68$. The equilibrium pressure at that composition is 60940 Pa. The equilibrium vapor pressures of the pure species at 40°C are :

$p^*_{CHCl_3}$ = 47705 Pa $\qquad p^*_{CH_3OH}$ = 35280 Pa

The activity coefficients using convention I are :

$\gamma^I_{CHCl_3} = \dfrac{60940}{47705} \simeq 1.28 \qquad \gamma^I_{CH_3OH} = \dfrac{60940}{35280} \simeq 1.73$

Using the Gibbs-Duhem relation (6.18), it can be shown that for a binary system (species A and B), the following relations hold :

$$\left.\begin{array}{r}x_A \left(\dfrac{\partial \mu_A}{\partial x_A}\right)_{T,p} = x_B \left(\dfrac{\partial \mu_B}{\partial x_B}\right)_{T,p} \\[2ex] x_A \left(\dfrac{\partial \ln \gamma_A}{\partial x_A}\right)_{T,p} = x_B \left(\dfrac{\partial \ln \gamma_B}{\partial x_B}\right)_{T,p}\end{array}\right\} \qquad (12.19)$$

Note that, if one of the species of the system behaves ideally, then we have :

$$\left(\dfrac{\partial \gamma_A}{\partial x_A}\right)_{T,p} = 0 \quad \Leftarrow \atop \Rightarrow \quad \left(\dfrac{\partial \gamma_B}{\partial x_B}\right)_{T,p} = 0 \qquad (12.20)$$

and the other one also behaves ideally.

Example

Consider a binary solution, for which the activity coefficients are given by :

$\ln \gamma^I_A = k(1-x_A)^2 \qquad \ln \gamma^I_B = k(1-x_B)^2$

Theses solutions are known as regular solutions. We can verify that these expressions fulfill the requirement of Eq. 12.19 :

$x_A \left(\dfrac{\partial \ln \gamma_A}{\partial x_A}\right)_{T,p} = -x_A\, k\, 2(1-x_A) = -2\, k\, x_A\, x_B$

$x_B \left(\dfrac{\partial \ln \gamma_B}{\partial x_B}\right)_{T,p} = -x_B\, k\, 2(1-x_B) = -2\, k\, x_A\, x_B$

12.5.2 Isobaric Diagram

We can also draw an isobaric equilibrium diagram, for a system containing the same chemical species as that of Fig. 12.1 (We selected a pressure = 0.1 bar). The vaporization curve and the condensation curve this time go through a *minimum*. The boiling temperature of the mixture at that composition is lower than that of either species when pure.

The domain where the system contains only liquid is below the vaporization curve. The domain where the system is only gaseous is above the condensation curve. For a point representative of the average composition at a temperature such that it is between these two curves, the liquid and vapor phase are at equilibrium. The composition of each of the phases corresponds to the intersection of a horizontal line with the corresponding curves. The compositions of the liquid and vapor phases are identical at the azeotropic composition. The azeotropic mixture behaves almost as a pure substance, since at a given pressure, it has a fixed boiling temperature and no concentration changes are observed by distillation.

At a different pressure, the azeotropic composition will be different (sometimes only slightly). For a pressure of 5 bar, the azeotrope corresponds to $x_{CHCl_3} \simeq 0.50$ at a temperature 105.63 °C.

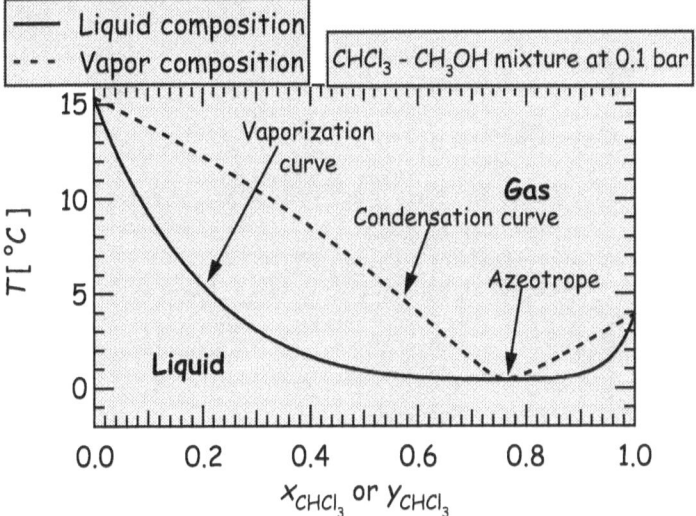

Figure 12.2 Isobaric diagram (0.1 bar) of a non-ideal solution, a mixture of chloroform and methanol with an azeotrope at $x_{CHCl_3} \simeq 0.76$ at a temperature of 0.43 °C.

In cases where the isobaric diagram presents a minimum as in Fig. 12.2, a fractional distillation may lead to the azeotropic mixture, while the liquid is enriched in one of the pure substances.

12.6 Standard State – Convention II for the Activity Coefficient

Some substances do not exist in the liquid state near the pressure or temperature of interest. They might be either solid or gas. It is thus impossible to prepare a solution containing a high mole fraction of such a substance. It is then appropriate and necessary to use a different convention for the activity coefficient and the reference state, known as convention II. The main species of the solution present at a high mole fraction is the **solvent**, liquid under the conditions of interest. The other species present at lower mole fractions in the solution are the **solutes**.

A solute behaves ideally when its mole fraction in a solution is small. These facts naturally lead us to the following conventions for the activity coefficients:

$$\left. \begin{array}{l} \mu_{solv} = \mu_{solv}^{*I}(T,p) + RT\ln\gamma_{solv}^{I} x_{solv} \\ \text{with } \gamma_{solv}^{I} \to 1 \text{ when } x_{solv} \to 1 \end{array} \right\} \text{for the solvent}$$

$$\left. \begin{array}{l} \mu_{solute} = \mu_{solute}^{*II}(T,p) + RT\ln\gamma_{solute}^{II} x_{solute} \\ \text{with } \gamma_{solute}^{II} \to 1 \text{ when } x_{solute} \to 0 \end{array} \right\} \text{for a solute}$$

(12.21)

Convention I is used for the solvent. At low solute concentrations, the solvent obeys Raoult's law.

The *chemical potential* $\mu_{solute}^{*II}(T,p)$ corresponds to the chemical potential of the *pure solute* ($x_{solute} = 1$) at T and p, behaving as an infinitely diluted species in the solvent ($\gamma_{solute}^{II} = 1$). This state has no real existence. For a solute i, using Eq. 12.21 in Eq. 12.3, we have:

$$\left. \begin{array}{l} p_i = \gamma_i^{II} x_i k_{Hi} \qquad k_{Hi} = \exp\left(\dfrac{\mu_i^{*II}(T,p) - \mu_i^{\ominus}(T)}{RT}\right) \\ \text{and} \\ \dfrac{p_i}{x_i} \to k_{Hi} \text{ when } x_i \to 0 \text{ and } \gamma_i^{II} \to 1 \end{array} \right\}$$

(12.22)

The constant k_{Hi} is **Henry's constant**. The fact that the partial pressure of a species, when it behaves ideally, is proportional to its

12 Non Ideal Solutions

mole fraction in the solution constitutes **Henry's law**. Note that Henry's constant is for a given solute (dilute species) in a given solvent. Note also that L'Hospital rule implies the following:

$$\frac{p_i}{x_i} \to \frac{dp_i}{dx_i} = k_{Hi} \quad \text{when } x_i \to 0 \qquad \left\{ \begin{array}{l} p_i \to 0 \\ \text{and} \\ \gamma_i^{II} \to 1 \end{array} \right\} \qquad (12.23)$$

Henry's constant also corresponds to the slope of the tangent to the curve representing the partial pressure of the solute. We placed in Fig. 12.1 the straight lines that correspond to Henry's law for dilute $CHCl_3$ in CH_3OH and dilute CH_3OH in CH_3Cl. Convention II can indeed be used to describe dilute solutions in system where it is possible to use convention I. In Fig. 12.1, for a system with a liquid composition corresponding to point X ($x_{CH_3OH} = 0.1$), the significance of convention II is illustrated. We have:

$$\gamma_{CH_3OH}^{II} = \frac{XR}{XH} < 1 \qquad (12.24)$$

The dependence on pressure of Henry's constant is small (derivation identical to that of Eq. 11.7). The pressure has very little effect on the chemical potential $\mu_i^{*II}(T, p)$. The **standard chemical potential** according to convention II is $\mu_i^{*II}(T, p^\ominus)$, the chemical potential of the species when it behaves as if it were pure but also infinitely dilute in the solvent, and at the standard state pressure, $p^\ominus$.

Example

For the chloroform methanol mixture at 40°C, Henry's constant for chloroform in methanol and methanol in chloroform are:

$$k_{H\ CHCl_3} = 132300 \text{ Pa} \qquad k_{H\ CH_3OH} = 261000 \text{ Pa}$$

For a dilute solution of CH_3OH in $CHCl_3$, we have:

$$p_{CH_3OH} \simeq x_{CH_3OH}\, k_{H\ CH_3OH} \quad \text{for example, for } x_{CH_3OH} = 0.01 \quad p_{CH_3OH} = 2610 \text{ Pa}$$

If the solution is not dilute then it is necessary to take into account the activity coefficient. Let us use convention II. The equilibrium pressure of a mixture for $x_{CH_3OH} = 0.1$ is 58580 Pa and the vapor composition is $y_{CH_3OH} = 0.241$:

$$\gamma_{CH_3OH}^{II} = \frac{p_{CH_3OH}}{x_{CH_3OH}\, k_{H\ CH_3OH}} = \frac{y_{CH_3OH}\, p_{eq}}{x_{CH_3OH}\, k_{H\ CH_3OH}} = \frac{0.241 \cdot 58580}{0.1 \cdot 261000} = 0.541$$

We can also obtain the activity coefficient using convention I, using Raoult's constant in place of Henry's constant. We obtain:

$$\gamma_{CH_3OH}^{I} = \frac{p_{CH_3OH}}{x_{CH_3OH}\, k_{R\ CH_3OH}} = \frac{y_{CH_3OH}\, p_{eq}}{x_{CH_3OH}\, k_{R\ CH_3OH}} = \frac{0.241 \cdot 58580}{0.1 \cdot 35280} = 4.00$$

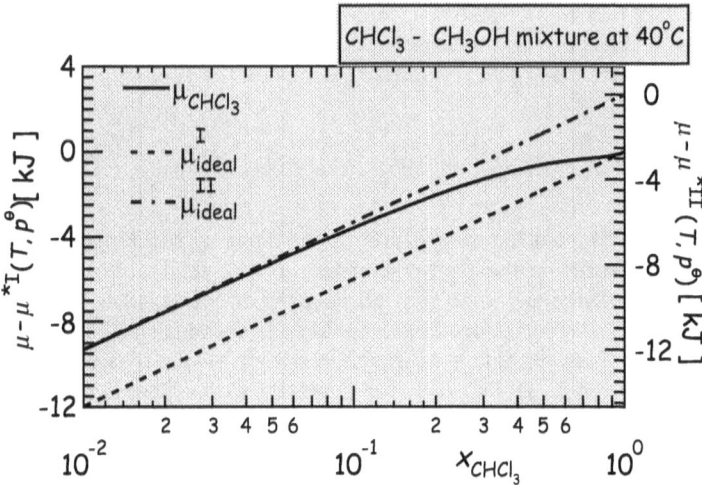

Figure 12.3 Chemical potential of $CHCl_3$ in a methanol chloroform mixture at 40°C. Ideal behaviors extrapolate to different values according to the convention selected.

Using Eq. 12.15 and Eq. 12.22, we obtain the relations that exist between the chemical potential of these two reference states and the two activity coefficients. We have:

$$\mu_i^{*II}(T,p) - \mu_i^{*I}(T,p) = R T \ln \frac{k_{Hi}}{k_{Ri}} \qquad (12.25)$$

$$\left. \begin{array}{l} p_i = \gamma_i^I x_i k_{Ri} \\ p_i = \gamma_i^{II} x_i k_{Hi} \end{array} \right\} \Rightarrow \gamma_i^I k_{Ri} = \gamma_i^{II} k_{Hi} \qquad (12.26)$$

In Fig. 12.3, we represent the change of the chemical potential of $CHCl_3$ with composition. The ideal behavior at low or high concentration is given by the dashed lines. The extrapolation to $x_{CHCl_3} = 1$ gives μ^{*I} (read on the left scale) for high $CHCl_3$ concentration and μ^{*II} (read on the right scale) for low $CHCl_3$ concentration. The standard chemical potentials difference at 40°C for $CHCl_3$ according to the convention selected is:

$$\left. \begin{array}{l} \mu_{CHCl_3}^{*II}(313.15, p^\circ) - \mu_{CHCl_3}^{*I}(313.15, p^\circ) = 8.3145 \cdot 313.15 \ln \frac{132300}{47705} \\ \simeq 2.66 \text{ kJ} \end{array} \right\} \qquad (12.27)$$

12.7 Liquid – Liquid Extraction

Let us illustrate the use of convention II for this type of process. We consider a system with two liquid phases α and β, containing a species i soluble in either phase. Let us characterize the equilibrium state for species i. At equilibrium, the chemical potential of species i is the same in both phases and we have :

$$\mu_i^\alpha = \mu_i^\beta \;\Rightarrow\; \mu_i^{\alpha*II} + RT \ln\left(\gamma_i^{\alpha II} x_i^\alpha\right) = \mu_i^{\beta*II} + RT \ln\left(\gamma_i^{\beta II} x_i^\beta\right) \tag{12.28}$$

The equilibrium condition 12.28 can also be written :

$$\left. \begin{array}{c} \dfrac{\gamma_i^{\beta II} x_i^\beta}{\gamma_i^{\alpha II} x_i^\alpha} = K = \dfrac{a_i^\beta}{a_i^\alpha} \\[1em] \text{with}\quad K = \exp\left(\dfrac{\mu_i^{\alpha*II} - \mu_i^{\beta*II}}{RT}\right) = \dfrac{K_H^\alpha}{K_H^\beta} \end{array} \right\} \tag{12.29}$$

The ratio of the activities of species i in each phase is a constant K known as the **partition coefficient**. We used Eq. 12.22 to obtain the partition coefficient in terms of Henry's constant for species i in each of the phases α and β.

12.8 Other Composition Scales and Standard States

12.8.1 Molality

Solution composition may also be expressed using the **molality scale**, m_i, instead of the mole fraction. The molality is the number of moles of solutes present in a kilogram of solvent. Let n_0 be the number of moles of solvent assumed to have a molar mass M_0 (kg mol^{-1})[†]. Molalities and mole fraction are related.

$$\left. \begin{array}{c} m_i = \dfrac{n_i}{n_0 M_0} \\[1em] x_i = \dfrac{n_i}{n_0 + \sum_j n_j} \end{array} \right\} \Rightarrow m_i = x_i \dfrac{n_0 + \sum_j n_j}{n_0 M_0} = x_i \left(\dfrac{1}{M_0} + \sum_j m_j\right) \tag{12.30}$$

The following relations are useful :

$$m_j M_0 = \dfrac{n_j}{n_0} \;\Rightarrow\; 1 + M_0 \sum_j m_j = 1 + \dfrac{1}{n_0}\sum_j n_j = \dfrac{1}{x_{solv}} \tag{12.31}$$

[†] Molar mass is distinct from molecular weight which is expressed in g mol^{-1}.

where x_{solv} is the mole fraction of solvent. We have :

$$x_i = \frac{m_i}{\left(\dfrac{1}{M_0} + \sum_j m_j\right)} \qquad m_i = \frac{x_i}{x_{solv} M_0} \qquad (12.32)$$

Molality and mole fraction are proportional in dilute solutions. Using this composition scale, the chemical potential of a solute is written :

$$\left.\begin{array}{l} \mu_i = \mu^*_{mi}(T, p) + RT \ln \gamma_{mi}\left(\dfrac{m_i}{m_i^\ominus}\right) = \mu^*_{mi}(T, p) + RT \ln \gamma_{mi} \, m_i \\[6pt] \quad = \mu^*_{mi}(T, p) + RT \ln a_{mi} \\[4pt] \text{with } \gamma_{mi} \to 1 \text{ when } m_i \to 0 \end{array}\right\}$$

(12.33)

where γ_{mi} is the activity coefficient of solute i in the molality scale, a_{mi} is the activity of species i in the molality scale. The use of the molality scale is indicated by the subscript m. The chemical potential of the pure solute $\mu^*_{mi}(T, p)$ when its activity is one, is here also not affected very much by pressure. The reference state thus selected corresponds to a fictitious state where the solute would behave ideally, $\gamma_{mi} = 1$ and molality is 1, $m_i^\ominus = 1$ mol kg^{-1}. Knowing that $m_i^\ominus$ is 1, it is often omitted in the expression of the chemical potential. The standard state corresponds to the pressure $p^\ominus$. For the solvent, convention I and the mole fraction scale are used.

Example

Consider a solution containing 0.1mol of CH_3OH in 0.9 mol of water. The mole fraction of the methanol in that solution is :

$$x_{CH_3OH} = 0.1$$

The molar mass of water is 18.015 10^{-3} kg mol^{-1}, the molality of this solution is :

$$m_{CH_3OH} = \frac{x_{CH_3OH}}{x_{H_2O} M_0} = \frac{0.1}{0.9 \cdot 18.015 \, 10^{-3}} = 6.168 \text{ mol kg}^{-1}$$

The chemical potential in the molality scale can be related to that in convention II by expressing the chemical potential using either scales. We get :

$$\left.\begin{array}{l} \mu_i = \mu_i^{*II}(T, p) + RT \ln \gamma_i^{II} x_i = \mu^*_{mi}(T, p) + RT \ln \gamma_{mi} \, m_i \\[4pt] \qquad \Downarrow \\[4pt] RT \ln \dfrac{\gamma_{mi} m_i}{\gamma_i^{II} x_i} = \mu_i^{*II}(T, p) - \mu^*_{mi}(T, p) \end{array}\right\}$$

(12.34)

12 Non Ideal Solutions

The second equation of 12.34 is also valid for a very dilute solution, where both activity coefficients have the value 1 and using the relation that exists between m_i and x_i for a dilute solution ($x_{solv} \approx 1$ in Eq. 12.29) leads to:

$$\mu_i^{*II}(T,p) - \mu_{mi}^{*}(T,p) \simeq R T \ln \frac{m_i}{x_i} = R T \ln \left(\frac{1}{M_0}\right) \quad (12.35)$$

The difference of the chemical potential in these two different reference states is independent of the solute and of pressure. Using the result of Eq. 12.35 in Eq. 12.34, we get the exact relation that exists between the two activity coefficients for these two standards.

$$\frac{\gamma_{mi}}{\gamma_i^{II}} = \frac{x_i}{m_i M_0} = x_{solv} \quad (12.36)$$

Example

When the solvent is water at 293.15 K, we have:

$$\mu_i^{*II}(T,p) - \mu_{mi}^{*}(T,p) = R T \ln\left(\frac{1}{M_0}\right)$$

$$= 8.3145 \cdot 293.15 \ln\left(\frac{1}{18.015 \cdot 10^{-3}}\right) = -9957 \text{ J mol}^{-1}$$

12.8.2 Concentration

The **concentration** is widely used as a composition scale. The following relations are relevant:

$$\left. \begin{array}{c} c_i = \dfrac{n_i}{V} = \dfrac{n_i}{n_0 \overline{V}_0 + \sum_i n_i \overline{V}_i} = \dfrac{x_i}{\langle V_m \rangle} \\[2ex] \text{with } \langle V_m \rangle = \dfrac{n_0 \overline{V}_0 + \sum_i n_i \overline{V}_i}{n_0 + \sum_i n_i} \end{array} \right\} \quad (12.37)$$

The concentration of species i depends on the average molar volume, $\langle V_m \rangle$, which, for a given system, depends on temperature and chemical composition. Concentrations are complicated functions of temperature and composition. However, for sufficiently dilute solutions, concentration and mole fractions are proportional, the average volume is then the volume of the solvent. The chemical potential of species i, is given by:

$$\mu_i = \mu^*_{c\,i}(T,p) + RT\ln \gamma_{c\,i} c_i = \mu^*_{c\,i}(T,p) + RT\ln a_{c\,i}$$
$$\text{with } \gamma_{c\,i} \to 1 \text{ when } c_i \to 0 \tag{12.38}$$

The official unit of concentration used is the **molarity** (1 mol dm^{-3}). The subscript c indicates that the composition scale used is concentration. The activity of species i in the concentration scale is $a_{c\,i}$. The reference state in the concentration scale corresponds to a fictitious state where the concentration of species i is $c_i^{\ominus} = 1$ mol dm^{-3} behaving ideally, $\gamma_{c\,i} = 1$ and usually the standard state pressure $p^{\ominus}$.

12.9 Law of Mass Action for Liquid Phase Systems

The law of mass action is obtained by using the appropriate expression for the chemical potential of the species present in solution. For a reaction written as:

$$\sum_i \nu_i M_i = 0 \tag{9.3}$$

The equilibrium state is obtained when $\Delta_r G$ is zero. The law of mass action is:

$$\Delta_r G = \sum_i \nu_i \mu_i = 0 = \sum_i \nu_i \left(\mu_i^*(T, p^{\ominus}) + RT\ln a_i \right)$$

$$0 = \sum_i \nu_i \mu_i^*(T, p^{\ominus}) + RT \sum_i \nu_i \ln a_i \tag{12.39}$$

$$0 = \sum_i \nu_i \mu_i^*(T, p^{\ominus}) + RT \ln \prod_i a_i^{\nu_i}$$

The subscripts x, m, or c should be added to activities according to the composition scale used. If the mole fraction scale is used, it would also be necessary to indicate whether convention I or II is used. A mix of composition scales can be used as long as the appropriate standard values of the chemical potentials are used when computing the equilibrium constant.

Keeping the general notation of Eq. 12.39, we can write:

$$\prod_i a_i^{\nu_i} = K_a$$
$$\text{with } RT\ln K_a = -\Delta_r G_T^{\ominus} = -\sum_i \nu_i \mu_i^*(T, p^{\ominus}) \tag{12.40}$$

The numerical value of the equilibrium constant depends on the standard states selected for the species present in the system. The activity coefficients or the activities of each species are also dependent on these choices.

Practically, convention I is always used for the solvent. For a solution where solutes are not too concentrated and when *the solvent takes part in the reaction*, it is conventional not to include it in the law of mass action. In this case, it is assumed that the error introduced by taking the solvent activity as equal to 1, ($\gamma_i \simeq 1$, $x_i \simeq 1$) is negligible.

The evaluations using the information contained in thermodynamic tables have to take into account the appropriate state of the species. Thus, for the solvent, the standard chemical potential will often be used if the pressure conditions are such that the chemical potential of the solvent is unaffected by the pressure conditions. As far as solutes are concerned, the standard chemical potential will be that corresponding to the selected composition scale, for example a molality of 1 if the molality scale is selected.

The selection of the standard state of the solutes participating in the reaction depends on the standard state usually adopted for each of them. Nothing precludes the use of convention I for example, if all of the species exist as liquid under the temperature and pressure conditions of the reaction.

Example

A system contains liquid water and neon gas at 298.15 K. The partial pressure of Ne is 1 atm (1.01325 bar). At equilibrium, the mole fraction of Ne in water is measured as $x_{Ne} = 8.152 \cdot 10^{-6}$. The equilibrium can be represented as :

$$Ne_{(g)} \rightleftharpoons Ne_{(aq)}$$

Using molality as the composition scale in the liquid and bar as the pressure unit in the gas phase, we can find the standard Gibbs energy of this reaction. Assuming that the activity coefficient is 1 for this low concentration, we have :

$$\Delta_r G^\ominus_{T,m} = -RT \ln \frac{\gamma_{mNe} x_{Ne}}{p_{Ne} M_0} = -8.3145 \cdot 298.15 \ln \frac{1 \cdot 8.152 \cdot 10^{-6}}{1.01325 \cdot 18.015} = 19122 \text{ J mol}^{-1}$$

We find in thermodynamics tables 19.3 kJ. We can also find the standard Gibbs free energy if convention II is used, assuming again that the activity coefficient is 1. We have :

$$\Delta_r G^{\ominus,II}_T = -RT \ln \frac{\gamma^{II}_{x\,Ne} x_{Ne}}{p_{Ne}} = -8.3145 \cdot 298.15 \ln \frac{1 \cdot 8.152 \cdot 10^{-6}}{1.01325} = 29079 \text{ J mol}^{-1}$$

The difference between these two values corresponds to the difference of the standard chemical potential for these two reference states for Ne in the liquid as obtained in the example of § 12.8.1.

12.10 Electrolytes

12.10.1 General Considerations

Some substances, when dissolved in an appropriate solvent, can dissociate as ions. The resulting solution is an **electrolyte**. The dissociation reaction is written as:

$$C_{\nu^+} A_{\nu^-} \rightleftharpoons \nu^+ C^{z^+} + \nu^- A^{z^-} \qquad (12.41)$$

where the charges on the ions, z_+ and z_- are measured in terms of the charge of a proton.

The solutions thus obtained are electrically neutral. The following relation holds:

$$\nu^+ z^+ + \nu^- z^- = 0 \qquad (12.42)$$

since the dissociation of a neutral entity produces the same number of positive and negative charges. The dissociation may be either partial (weak electrolytes) or total (strong electrolytes). Molality is the composition scale usually used for electrolytes. The electroneutrality is a property of the solution that leads to a relation between the molalities[†] of the ions (intensive variables) and is written:

$$\left. \sum_i z_i^+ m_{i+} + \sum_j z_j^- m_{j-} = 0 \right\} \qquad (12.43)$$

$i \Leftrightarrow$ cations $\quad j \Leftrightarrow$ anions

The existence of this relation must be taken into account when determining the variance of the system.

The method to compute the variance of an electrolyte solution can be illustrated by the very simple case of pure water in equilibrium with its vapor. So far, we considered that the system contained a single chemical species H_2O to find (using 10.41):

$$v = n + 2 - \varphi - r - s = 1 + 2 - 2 - 0 - 0 = 1 \qquad (12.44)$$

The system is univariant. For a system containing pure water, at some selected temperature, the pressure of the liquid in equilibrium with its vapor is the vapor pressure of water.

We can refine our vision of such a system by considering that the liquid phase also contains the ions H^+ and OH^-. The number of species in the system is now three ($c = 3$). We must now take into account the dissociation reaction that is the only independent reaction possible between these species ($r = 1$) and the electroneutrality relation ($s = 1$). Proceeding in this fashion, the variance of the system is found to be:

[†] or mole fractions, concentrations according to the composition variable selected.

$$v = n + 2 - \varphi - r - s = 3 + 2 - 2 - 1 - 1 = 1 \quad (12.45)$$

The two approaches provide, as could be anticipated, the same result. A simplified description of systems is often quite adequate to find their variance.

12.10.2 Chemical Potential of Ions in Solution

The chemical potential of ions is expressed using the same formal mathematical expression as for solutes. The chemical potential of an ionic species i is given by either of the two following expressions:

$$\mu_i = \mu^*_{m\,i}(T, p) + R T \ln \gamma_{m\,i}\, m_i = \mu^*_{m\,i}(T, p) + R T \ln a_{m\,i} \quad (12.46)$$

The activity coefficient describes the difference between the actual chemical potential of the ion and what it would be if the ion behaved ideally. The behavior of ions is strongly influenced by the coulombic interactions, which are quite large and are the main source of the non-ideal behavior of electrolytes, even at low concentrations. Each ion is surrounded by the ionic atmosphere of the other ions present. The **Debye Hückel theory** (approximate but simple) leads to a possible estimate of the activity coefficient of an ion in a solvent where electrolyte concentrations are sufficiently small. The theory involves the ionic strength of the solution defined by (on a molality basis):

$$I_m = \frac{1}{2} \sum_i m_i z_i^2 \quad (12.47)$$

The activity coefficient of an ion tends towards 1 in a solution of ionic strength 0. An ion behaves almost ideally in a solution of sufficiently weak ionic strength.

> The standard state of an ion is a state of molality 1 where **all ions** behave as if the solution were extremely dilute at the standard state pressure.

Example

Consider a solution where 0.1 mol of K_2SO_4 is dissolved in 55.5 mol of water. The molality of the solution is:

$$\frac{x_{K_2SO_4}}{x_{H_2O}} = \frac{n_{K_2SO_4}}{n_{H_2O}} \Rightarrow m_{K_2SO_4} = \frac{x_{K_2SO_4}}{x_{H_2O} M_0} = \frac{0.1}{55.5 \cdot 18.015\ 10^{-3}} \approx 100 \text{ mol kg}^{-1}$$

This strong electrolyte completely dissociates and the molalities of each type of ion present in the solution are related to the molality in the salt. The ionic strength is obtained as:

$$\left. \begin{array}{l} m_{K^+} = 0.2 \text{ mol kg}^{-1} \\ m_{SO_4^{2-}} = 0.1 \text{ mol kg}^{-1} \end{array} \right\} \Rightarrow I_m = \frac{1}{2}(0.2 \cdot 1^2 + 0.1 \cdot 2^2) = 0.3 \text{ mol kg}^{-1}$$

12.10.3 Dissociation Equilibrium

Let us write the expression for the differential of the Gibbs energy of a solution containing the solvent, the species $C_{\nu_+}A_{\nu_-}$ undissociated (subscript $_{un}$), as well as the ions in which it dissociates. We have:

$$dG = V\,dp - S\,dT + \mu_{solv}\,dn_{solv} + \mu_{un}\,dm_{un} + \mu_+\,dm_+ + \mu_-\,dm_- \quad (12.48)$$

The chemical potential of each type of ion present is given by:

$$\left.\begin{array}{l} \mu_+ = \left(\dfrac{\partial G}{\partial m_+}\right)_{p,\,T,\,n_{solv},\,m_{un},\,m_-} \\[1em] \mu_- = \left(\dfrac{\partial G}{\partial m_-}\right)_{p,\,T,\,n_{solv},\,m_{un},\,m_+} \end{array}\right\} \quad (12.49)$$

These potentials do not correspond to a physically realistic property since it is not possible to vary the concentration of one ion while keeping the concentration of the other one constant, the electroneutrality of the solution would not be achieved. Using the stoichiometric coefficients of the dissociation equilibrium (reaction 12.41), we can write the differential of the dissociation reaction extent, $d\xi$:

$$-dm_{un} = \frac{dm_+}{\nu^+} = \frac{dm_-}{\nu^-} = d\xi \quad (12.50)$$

For an *isothermal and isobaric* system, where the amount of solvent is constant, we have the equilibrium condition:

$$dG = (-\mu_{un} + \nu^+\mu_+ + \nu^-\mu_-)\,d\xi \Rightarrow \begin{cases} -\mu_{un} + \nu^+\mu_+ + \nu^-\mu_- = 0 \\ \text{at equilibrium} \end{cases} \quad (12.51)$$

Thus the equilibrium condition for the ionic dissociation reaction is similar to the equilibrium condition for any chemical reaction. The law of mass action can be applied to the dissociation reaction.

The equilibrium condition can be written for a reaction involving ions, using expression 12.46 for the expression of the chemical potential. We obtain the expression of the law of mass action under a form identical to Eq. 12.40.

$$\left.\begin{array}{c} \displaystyle\prod_i a_i^{\nu_i} = K_a \\[1em] \text{with} \quad R\,T\ln K_a = -\Delta_r G_T^\ominus = -\displaystyle\sum_i \nu_i\,\mu_i^*(T, p^\ominus) \end{array}\right\} \quad (12.52)$$

Example

Consider the reaction:

$$H_2O\,(l) \rightleftharpoons H^+ + OH^-$$

12 Non Ideal Solutions

The law of mass action for this equilibrium is :

$$\frac{a_{H^+} \cdot a_{OH^-}}{a_{H_2O}} = K_a$$

We have seen (§10.4.1) that the activity of water can be here taken as 1. We have the following data found in tables. The data for the ions are given at 298.15 K for use with the molality scale :

Species	$\Delta_f G^\ominus_{298.15}$ [kJ mol^{-1}]
$H_2O_{(l)}$	– 237.129
H^+	0
OH^-	– 157.244

We find the standard Gibbs energy of reaction to be :

$$\Delta_r G^\ominus_{T,m} = 1 \cdot (-157.244) - 1 \cdot (-237.129) = 79.885 \text{ kJ mol}^{-1} = 79885 \text{ J mol}^{-1}$$

The equilibrium constant is :

$$K_a = \exp\left(-\frac{\Delta_r G^\ominus_T}{RT}\right) = \exp\left(-\frac{79885}{8.3145 \cdot 298.15}\right) = 1.01 \, 10^{-14}$$

For pure water, the amounts of H^+ and OH^- are quite small and the ion activities are practically equal to their concentration. We obtain the well known result at 298.15 K :

$$[H^+][OH^-] = 1.01 \, 10^{-14}$$

12.10.4   Hydrogen Ion Convention for Aqueous Solutions

Measuring thermodynamic properties of a single type of ion is impossible since one cannot prepare solutions containing only one type of ions.

Tables of the formation properties of ions can however be prepared if an arbitrary value for the Gibbs energy of formation is selected for one type of ion at each temperature and for each solvent. This method provides a reference with respect to which the Gibbs energy of formation of the other individual ions can be determined. By convention, the Gibbs energy of formation of the hydrogen ion is zero in an aqueous solution in its standard state (molality 1, ideal behavior, pressure $p^\ominus$) from gaseous hydrogen at the standard state pressure $p^\ominus$. We have :

$$\left\{ \begin{array}{l} \frac{1}{2} H_2 (g) \rightleftarrows H^+ (aq) + e^- \\[4pt] \Delta_r G^\ominus_T = 0 \\ \quad = \Delta_f G^\ominus_T(H^+ (aq)) + \Delta_f G^\ominus_T(e^-) - \frac{1}{2}\Delta_f G^\ominus_T(H_2 (g)) \\ \text{at any temperature} \end{array} \right. \qquad (12.53)$$

Since by convention, we have :

$$\left.\begin{array}{l} \Delta_f G^\ominus_T(H^+(aq)) = 0 \\ \Delta_f G^\ominus_T(H_2(g)) = 0 \end{array}\right\} \Rightarrow \Delta_f G^\ominus_T(e^-) = 0 \text{ at any temperature} \quad (12.54)$$

With the selected convention, the Gibbs energy of formation of the electron in its "standard state" is also zero. Using the Gibbs-Helmholtz equation (5.25), we find that the enthalpy of reaction 12.53 is zero.

$$\left.\begin{array}{c} \dfrac{\partial}{\partial T}\left[\dfrac{\Delta_r G^\ominus_T}{T}\right] = -\dfrac{\Delta_r H^\ominus_T}{T^2} = 0 \\ \Downarrow \\ \Delta_r H^\ominus_T = \Delta_f H^\ominus_T(H^+(aq)) + \Delta_f H^\ominus_T(e^-) - \dfrac{1}{2}\Delta_f H^\ominus_T(H_2(g)) = 0 \end{array}\right\} \quad (12.55)$$

By applying the Gibbs-Helmholtz equation to 12.55 and with the result of 12.54, we find that the enthalpy of formation of the hydrogen ion and of the electron in their standard state are both zero at all temperatures.

$$\left.\begin{array}{c} \dfrac{\partial}{\partial T}\left[\dfrac{\Delta_f G^\ominus_T(H^+(aq))}{T}\right] = -\dfrac{\Delta_f H^\ominus_T(H^+(aq))}{T^2} = 0 \\ \Downarrow \\ \Delta_f H^\ominus_T(H^+(aq)) = 0 \Rightarrow \Delta_f H^\ominus_T(e^-) = 0 \text{ at any temperature} \end{array}\right\} \quad (12.56)$$

From these results we obtain the entropy of reaction 12.53. We have :

$$\left.\begin{array}{c} \Delta_r S^\ominus_T = \dfrac{\Delta_r H^\ominus_T - \Delta_r G^\ominus_T}{T} = 0 = S^\ominus_T(H^+(aq)) + S^\ominus_T(e^-) - \dfrac{1}{2}S^\ominus_T(H_2(g)) \\ S^\ominus_T(H^+(aq)) = 0 \Rightarrow S^\ominus_T(e^-) = \dfrac{1}{2}S^\ominus_T(H_2(g)) \end{array}\right\}$$
(12.57)

By convention, the standard entropy of the hydrogen ion is taken to be zero at all temperatures, which allows the determination of the standard entropy of other ions. The value of the standard entropy of the electron obtained here has absolutely no thermodynamic significance, but is needed to carry out the evaluations.
Table 12.1 give a summary of the results obtained.

12 Non Ideal Solutions   175

Species	$\Delta_f H^\ominus_{298.15}$ (kJ mol^{-1})	$\Delta_f G^\ominus_{298.15}$ (kJ mol^{-1})	$S^\ominus_{298.15}$ (J mol^{-1} K^{-1})
(e$^-$)	0	0	65.342
H$^+$ (aq)	0	0	0
H$_2$ (g)	0	0	130.684

Table 12.1 Standard properties used for and derived from the hydrogen ion convention.

12.10.5 Electrode Potential

An electrochemical reaction taking place on an electrode is conventionally written in the direction where reduction takes place (not oxidation as we wrote for reaction 12.53). The **half cell standard potentials** are listed in electrochemical data tables. For a species, this potential is known as the **standard reduction potential** or the **standard redox potential** and it is obtained by :

$$E^\ominus_T = -\frac{\Delta_r G^\ominus_T}{nF} \qquad \text{Oxidant} + n\,e^- \rightleftharpoons \text{Reductant}$$

$$F = 96485.3415\ C\ mol^{-1}$$

(12.58)

where n is the stoichiometric coefficient affected to the electron in the oxred reaction, F is the Faraday constant. With this convention, the more oxidant a species is, the higher its electrode potential.

The reactions taking place in an electrochemical cell are often represented by writing that the oxidation reaction takes place on the left electrode (anode, negative pole) and the reduction reaction on the right electrode (cathode, positive pole). The standard potential of a cell is obtained by subtraction of the standard potential of the oxidation reaction from the standard potential of the reduction reaction.

Using Eq. 9.34 and 12.58, we can find the standard entropy of reaction as a function of the standard electrode potential. All the quantities being standard, the pressure $p^\ominus$ is constant, we get :

$$\Delta_r S^\ominus_T = -\frac{d\Delta_r G^\ominus_T}{dT} = nF\frac{dE^\ominus_T}{dT} \qquad (12.59)$$

The variation of the electrode potential with temperature provides a means of measuring entropies of reactions. The measurements of cell potentials lead to Gibbs energies of reaction and entropies of reaction.

Example

For the fuel cell mentioned in the example of § 9.7.2 at 298.15 K, we just saw that:

$$2\,e^- + 2\,H^+ \longleftarrow H_2(g) \quad \Delta_r G_T^\ominus = 0 \quad \Rightarrow \quad E_T^\ominus = 0$$

For the other electrode reaction:

$$\tfrac{1}{2}\,O_2(g) + 2\,H^+ + 2\,e^- \longrightarrow H_2O(l)$$

we have, using the Gibbs energy of formation of water found in thermodynamic tables:

$$\Delta_r G_T^\ominus = \Delta_f G_T^\ominus (H_2O) - \tfrac{1}{2}\Delta_f G_T^\ominus (O_2) - 2\,\Delta_f G_T^\ominus (H^+) - 2\,\Delta_f G_T^\ominus (e^-)$$

$$= -237.129 \text{ kJ}$$

The Gibbs energies of formation of the other species in the equation are zero. The standard redox potential for the reduction of oxygen is (using J as the energy unit):

$$E_T^\ominus = -\frac{\Delta_r G_T^\ominus}{n\,F} = -\frac{-237.129 \cdot 10^3}{2 \cdot 9.6485 \cdot 10^4} \simeq 1.229 \text{ V}$$

13. Bibliography

13.1 Textbooks

The Bases of Chemical Thermodynamics, volumes 1 & 2, Michael Graetzel, Pierre Infelta, Universal Publisher, UPublish.com, Parkland, Florida, 2000 (Revised printing 2002).
Fundamentals of Statistical and Thermal Physics, Frederic Reif, International edition, McGraw-Hill, 1985.
Physical Chemistry, Robert A. Alberty, Robert J. Silbey, 1st edition, John Wiley & Sons, 1992.
The Principles of Chemical Equilibrium, Kenneth Denbigh, 4th edition, Cambridge University Press, 1981.
The Virial Coefficients of Pure Gases and Mixtures, J. H. Dymond and E. B. Smith, Clarendon Press, 1980.

13.2 Handbooks and Tables

Atomic Energy States, Natl. Bur. Standards, Circ, 1, 1949, C. E. Moore
CRC Handbook of Chemistry and Physics, 83rd edition, David R. Lide, 2002.
JANAF Thermochemical Tables, M. W. Chase Jr. et al., 3rd edition, New York, American Institute of Physics, 1985, Journal of Physical and Chemical Reference Data, 14, 1985 supp. 1.
J. Phys. Chem. Ref. Data 1 (1972) 221-277, E. S. Domalski, Selected values of Heats of Combustion and Heats of Formation of Organic Compounds.
J. Phys. Chem. Ref. Data 11 (1982) Supplement No. 2, Donald D. Wagman et al. , The NBS Tables of Chemical Thermodynamic Properties.
Quantities, Units and Symbols in Physical Chemistry, prepared by I. Mills, T. Cvitaš, K. Homann, N. Kallay, K. Kuchitsu, 2nd edition, 1993.

The Properties of Gases and Liquids, Robert C. Reid, John M. Prausnitz, Bruce E. Poling, 5th edition, McGraw-Hill Book Company, New York, 2000.
Thermodynamic and Physical Property Data, Carl L. Yaws, Gulf Publishing Company, 1992.
Vapor-Liquid Equilibria, M. Hirata, S. Ohe, K. Nagama, Kodansha limited Elsevier Scientific Publishing Co., 1975.

13.3 Articles

"The Second Law: Statement and Applications", P. Infelta, Journal of Chemical Education **79**, 884, 2002.

"Conversion of Standard State Thermodynamic Data to the New Standard State Pressure", R. D. Freeman, Journal of Chemical Education **62**, 681,1985.

Index

Activity
    coefficient and molality 166
    coefficient in the concentration scale 168
    coefficient of a species in a real solution 153
    coefficient using convention I 157
    coefficient using convention II 162
    effect of pressure and temperature on the — coefficient 155
    of a pure condensed phase 125
    of a species in a real solution 153
Adiabatic
    chemical equilibrium of an — isobaric system 121
    chemical equilibrium of an — isochoric system 122
    closed — system 20, 53, 55
    enclosure 2, 16
    process 16
    reversible process 17
    reversible process of an ideal gas 29
Auxiliary
    functions 39
    state variable 3
Azeotropy 158-162
Beau de Rochas cycle 34
Binary
    mixture 139, 144
    system 158
Boiling point
    — 141, 144
    elevation of the — of a solvent in the presence of a solute 150
Bubble point 141
Carnot cycle
    — 17, 30
    reverse — 18

Change
    isothermal reversible — of a gas 11
    of a state variable 5
    of the Gibbs energy in a mixture of chemical species 117
    of the Gibbs energy with the extent of reaction 117, 118
Characteristic variables 42-46
Châtelier
    Le —'s principle 135
Chemical
    conversion of — energy into work 113, 114
    independent — reactions 128
    reaction 103
Chemical equilibrium
    effect of an inert gas on a — 137
    effect of pressure on a — 136
    effect of temperature on a — 136
    effect of volume on a — 137
    of an adiabatic isobaric system 121
    of an adiabatic isochoric system 122
    of an isobaric isothermal system 119
    of an isothermal isochoric system 120
Chemical potential
    definition 43
    of a multiphase system at equilibrium 90
    of a pure condensed phase 125
    of a pure ideal gas 69
    of a pure liquid 140
    of a pure real gas 76
    of a real gas in a mixture 83
    of a species in a real solution 153
    of an ideal gas in an ideal gas mixture 73
    of an ideal species 139
    of ions in solution 171
    reference state — 140, 157, 162, 166, 168
    standard — 69, 157, 162, 166, 168

I

Chemical reaction
    energetics of —s 103
    isothermal and isobaric — 104
    system with several —s 131
Clapeyron equations 92
Clausius inequality 22
Clausius-Clapeyron equations 94
Closed
    adiabatic system 20
    system 1, 40-41
    system with one chemical reaction, 103
Coefficient
    activity — and molality 166
    activity — and the concentration scale 168
    fugacity — of a pure real gas 77
    fugacity — of a real gas in a mixture 83
    isobaric — of thermal expansion 4
    isothermal compressibility — 4
    Joule-Thomson — 82
    of performance 32
    partition — 165
    stoichiometric —s 103
    thermal expansion — 4
    van der Waals —s 79
Combustion
    internal — engine 34
Components
    number of — in a system 133
Composition
    concentration scale, 167
    molality scale, 165
    mole fraction scale, 61, 72, 140, 153
    of the vapor phase of a binary solution 142, 144, 159, 161
Compressibility
    factor 77, 81
    isothermal — coefficient 4
Compression
    isothermal — of an ideal gas 26
    ratio 35

Concentration
    and composition 167
    scale 167
Condensation
    — 78
    curve 142, 144, 158, 161
Condensed phase
    chemical potential of a pure— 125
Constant
    Avogadro's — 13
    critical —s 79
    Henry's — 163
    Raoult's — 141
    standard equilibrium — 123
    thermodynamic equilibrium — 123, 127
    volume thermal expansion coefficient 4
Constant pressure
    process at — 119, 121
Constant volume
    process at — 40, 120, 122
Continuity of the fluid state 99
Convention
    for aqueous systems 173
    I for the standard state and activity coefficient 157
    II for the standard state and activity coefficient 162
    relation between — I and II activity coefficients 164
    relation between — II and the molality scale 166
    sign — for energy exchanges 9
    —s for activity coefficients in solutions 153
Conversion of chemical energy into work 113
Corresponding states of real gases 80
Critical
    constants 79
    point 79
    temperature 78
Curve
    condensation — 142, 144, 158, 161
    dew point — 142, 144

Index

fusion — 98
inversion — 82
sublimation — 98
vaporization — 98, 141, 144, 161
Cycle
   Beau de Rochas — 34
   Carnot — 17, 30
   dithermal — 17
   Otto — 34
   reverse Carnot — 18
   Stirling — 36
Cyclic process 5, 16
Dalton's law 72
Debye-Hückel theory 171
Degrees of freedom
   number of — 91, 133
Delay
   phase change —s 99
Dew point curve 142, 144
Diagram
   isobaric — for a perfect binary solution 143
   isothermal — for a perfect binary solution 141
   phase — 98
Diathermal enclosure 2
Differential
   exact — 6
Differential expression
   for the enthalpy 43
   for the free energy 44
   for the free enthalpy 44
   for the Gibbs energy 44
   for the Gibbs function 44
   for the Helmholtz energy 44
   for the Helmholtz function 44
   for the internal energy 42
   of state functions (multiple phases) 89
   —s for open systems, 45, 132
Dilute solutions 157, 166, 167
Dissociation equilibrium 172
Dithermal cycle 17

Effect
   Joule-Thomson — 81
   of an inert gas on chemical equilibria 137
   of pressure and temperature on the activity coefficient 155
   of pressure and volume on entropy at 0 K 101
   of pressure on chemical equilibria 136
   of pressure on phase equilibrium 92
   of temperature on chemical equilibria 136
   of temperature on the entropy of reaction 112
   of temperature on the Gibbs energy of reaction 112
   of temperature on the latent heat 97
   of temperature on vapor pressure 97
   of volume on chemical equilibria 137
Efficiency of an engine 31, 32
Electrical work 12
Electrode potential
   and standard entropy of reaction 175
   definition 175
Electrolyte
   chemical potential of ions in solution 171
   dissociation 170
Electroneutrality of a solution 170
Elevation of the boiling temperature of a solvent in the presence of a solute 150
Enclosure
   adiabatic — 2, 16
   diathermal — 2
Endothermic reaction 109
Energetics of chemical reactions 103
Energy
   free — 44
   Gibbs — 39
   Helmholtz — 39, 44
   internal — 15, 42
   sign convention for — exchanges 9
   various forms of — 13

Enthalpy
— 3, 39
explicit expression for the — 62
of an ideal gas mixture 71
of mixing 74, 84, 146
of phase change 94
of reaction 105
standard — of formation 107
Entropy
at 0 K 101
definition 19
effect of temperature on the — of reaction 112
evaluating — 100
global — 27, 54
of an ideal gas 70
of mixing 74, 146, 155
of reaction 105
standard — 107
Equation
Clapeyron —s 92
Clausius-Clapeyron —s 94
Gibbs-Duhem — 62
Kirchhoff's — 110
van der Waals — of state 78
virial — 77
Equation of state
of an ideal gas 50
thermodynamic — 49
van der Waals — 78
Equilibria
simultaneous — 132
Equilibrium
chemical — displacement laws 135
condition 24, 57
dissociation — 172
effect of an inert gas on a chemical — 137
effect of pressure and temperature on liquid vapor — 146
effect of pressure on a chemical — 136
effect of pressure on phase — 92
effect of temperature on a chemical — 136
effect of volume on a chemical — 137
gas and condensed phase — 94
liquid vapor — 141
of a chemical system 1, 117
of two phases 92
system in a state of — 1
thermal — 1, 2, 24
Equilibrium constant
— 123
effect of temperature on the — 134
in heterogeneous systems 126
in solutions 169
in the gas phase 124
in the liquid phase 169
standard — 123
thermodynamic — 123, 127
Euler's identity 59
Eutectic 149
Evaluation of entropies 100
Exact differential 6
Excess
entropy of mixing 155
entropy of mixing (gases) 86
Gibbs energy of mixing 154
Gibbs energy of mixing (gases) 85
variable of mixing (gases) 85
variables of mixing 154
Exothermic reaction 109
Expansion
isobaric coefficient of thermal — 4
isothermal — of an ideal gas 26
monothermal — of an ideal gas 27
Expression
explicit — for enthalpy 62
explicit — for the Gibbs energy 62
explicit — for the Helmholtz energy 62
explicit — for the internal energy 62
of the molar entropy of an ideal gas 70
Extensive
properties of — variables, 60
variable 2, 39

Extent of reaction
  — 103
  and expression of enthalpy 110
  and work 113, 114
  change of the Gibbs energy with the — 117, 118
Extraction
  liquid-liquid — 165
Field
  magnetic — 13
First law of thermodynamics 15
Fluid
  continuity of the — state 99
Force
  external — 9
  van der Waals —s 75
  work of an external — 9
Formation
  standard enthalpy of — 107
  standard Gibbs energy of — 107
  standard variables of — 107
Free
  energy 3, 39, 44
  enthalpy 3, 39
Freezing point lowering by a solute 147
Fuel cell 114
Fugacity
  and the law of mass action 123
  coefficient of a pure real gas 77
  definition 76
  of a real gas in a mixture 83
Function
  auxiliary state — 39
  homogeneous —s 59
  properties of homogeneous —s 59
  state — 2, 89
Fusion 98
Gas
  constant 8
  effect of an inert — on vapor pressure 95
  equation of state of an ideal — 50

ideal — mixture 70
  real —es 75
  thermodynamics of —es 69
General process 5, 16
Generalization of Hess' law 108
Gibbs energy
  — 3, 39, 55
  and extent of reaction 117
  at 0 K 102
  effect of temperature on the — energy 112
  excess — of mixing 154
  excess — of mixing (gases) 85
  explicit expression for the — 62
  of a mixture of reactants 117
  of formation of hydrogen ions 173
  of mixing 74, 146, 154
  of mixing in solutions 146
  of mixing of real gases 85
  of reaction 104, 105
Gibbs function 3, 39
Gibbs-Duhem equation 62
Gibbs-Helmholtz equations 46
Global entropy
  — 27, 54
  and spontaneity 27
Half-cell standard potential 175
Heat
  — 15
  engine 17, 31
  pump 31, 34
  source 1, 17, 53
  system in contact with one — source 17, 53, 55, 119, 120
  system in contact with several — sources 22
  system in contact with two — sources 17
  transfer 24
Heat capacity
  at constant pressure 41
  at constant volume 40
  molar — 29
  of reaction 105

Helmholtz energy
— 3, 39, 44, 53
at 0 K 102
explicit expression for the — 62
of reaction 105
Helmholtz function 3, 39, 44
Henry's
constant 163
law 163
Hess' law 108
Heterogeneous system
— 1
—s and the law of mass action 126
Homogeneous
functions 59
system 1
Hydrogen ion convention for aqueous solutions 173
Ideal
gas mixture 70
mixture of gases 86
perfect and — solutions 139-152
species 140
Ideal gas
Chemical potential of a pure — 69
definition 50
law of mass action for —es 124
mixture of —es 70
partial pressure of an — 71
Identity
Euler's — 59
Implications
of the first law 16
of the second law 20
of the third law 101
Independent reactions
— 128
and the law of mass action 132
number of — 128
Inert
effect of an — gas on chemical equilibria 137

Infinitesimal process 16
Intensive
properties of — variables 60
variable 2, 39, 45, 61
Internal energy
— 15, 42
differential expression for the — 23
explicit expression for the — 62
expression for the — 23
of an ideal gas mixture 70
of mixing 85
of reaction 105
Inversion curve and temperature 82
Ionic strength 171
Ions
chemical potential of — in solutions 171
law of mass action for — 172
Irreversible
(real) process 7
process 19
Isenthalpic process 82
Isentropic process 20, 29
Isobaric
chemical equilibrium of an adiabatic — system 121
coefficient of thermal expansion 4
diagram 161
diagram for a perfect binary solution 143
equilibrium of an — isothermal system 119
expansivity 4
process 40
Isochoric
chemical equilibrium of an adiabatic — system 122
equilibrium of an isothermal — system 120
Isochoric process 40, 120, 122
Isolated system 1
Isothermal
compressibility coefficient 4
compression of an ideal gas 26
diagram 141, 158

Index

equilibrium of an isobaric — system 119
equilibrium of an isochoric — system 120
expansion of an ideal gas 26
representation of perfect solution liquid vapor equilibrium 141
reversible process 17

Joule
  energy unit 15
  the — experiment 15

Joule-Thomson
  coefficient 82
  effect 81

Kirchhoff's equation 110

Latent heat
  effect of temperature on the — 97
  of phase change 94

Law
  Dalton's — 72
  displacement —s of equilibria 135
  Henry's — 163
  Hess' — 108
  Raoult's — 141

Law of mass action
  — 123
  for a heterogeneous system 126
  for ideal gases 124
  for ions 172
  in a liquid phase 168
  in solutions 168

Law of thermodynamics
  first — 15
  second — 17
  third — 101
  zeroth — 2

Le Châtelier's principle 135

Lennard-Jones interaction 75

Lever rule 143, 144, 159

Lewis-Randall rule 87

Lowering of the freezing point in the presence of a solute 147

Magnetic field 13

Mass action
  law of — 123
  law of — for a heterogeneous system 126

Maximum
  amount of work 55, 56, 113
  azeotrope 158
  entropy 24
  extent of reaction 117

Maxwell's relations 46-48

Measurement of partial molar volumes 67

Mechanical equilibrium 1, 89

Mixing
  enthalpy of — 74, 84, 146
  entropy of — 74, 146
  excess entropy of — 155
  excess variable of — (gases) 85
  Gibbs energy of — 74, 146
  Internal energy of — 85
  irreversible — of two ideal gases 75
  properties of ideal solutions 145
  variables of — 154
  variables of — for ideal gases 74
  variables of — for real gases 84
  volume of — 145
  volume of — of real gases 84

Mixture
  binary 139, 144
  chemical potential of an ideal gas in an ideal gas — 73
  fugacity of a real gas in a — 83
  ideal gas — 70
  ideal — of gases 86
  law of mass action for an ideal gas — 123
  —s of ideal gases 70
  —s of real gases 83

Molality
  definition 165
  scale 165

Molar
  heat capacity 29
  partial — enthalpy of an ideal gas 73
  partial — entropy in a solution 155
  partial — entropy of an ideal gas 73

partial — quantities 63, 67
partial — volume of an ideal gas 72
properties 65
quantities 65
Molarity 168
Mole fraction 61, 72, 140
Molecular interactions in real gases 75
Monobaric
    and monothermal process 56
    process 41
Monothermal
    and monobaric process 56
    process 17
Natural variables 42
Number
    of chemical species in a system 91, 133
    of components of a system 133
    of independent reactions 128
    of independent species in a system 133
    of phases in a system 91
Open system 42, 90
Osmotic pressure 151
Otto cycle 34
Partial molar
    enthalpy of an ideal gas 73
    entropy in a solution 155
    entropy of an ideal gas 73
    quantities 63, 67
    quantities of pure substances 65
    volume of an ideal gas 72
Partial pressure of an ideal gas 71
Partition coefficient 165
Perfect
    and ideal solutions 139-152
    solution 141
Phase
    change delays 99
    chemical potential of a condensed — 125
    definition 89
    diagram 98
    law of mass action in the liquid — 168

number of —s in a system 91
rule 3, 91, 133
rule and electrolytes 170
Physical meaning of
    the Gibbs function 55
    the Helmholtz function 53
Point
    critical — 79
    triple — 19, 99
Potential
    half-cell standard — 175
    reduction — 175
Pressure
    effect of — on chemical equilibria 136
    effect of — on the activity coefficient 155
    intensive variable 61
    osmotic — 151
    standard state — 69, 105
Principle
    Le Châtelier's — 135
Process
    adiabatic — 16
    at constant external pressure 11, 41, 56
    at constant pressure 119, 121
    at constant volume 40, 120, 122
    chemical — 103
    cyclic — 5, 16
    irreversible — 7, 19
    isenthalpic — 82
    isentropic — 20, 29
    isobaric — 40
    isochoric — 40, 120, 122
    monobaric — 41
    monothermal and monobaric — 56
    monothermal — 17
    reverse — and work 12
    reversible adiabatic — of an ideal gas 29
    reversible — 7, 19
    spontaneous — 75

Index

Properties
  of extensive variables 60
  of homogeneous functions 59
  of intensive variables 60
Pure
  chemical potential of a — ideal gas 69
  chemical potential of a — real gas 76
  real gas 75
Rank of a matrix 129
Raoult's
  constant 141
  law 141
Reaction
  endothermic — 109
  exothermic — 109
  extent of — 103
  independent —s 128
Real
  mixtures of — gases 83
  solution 153
  volume of mixing of — gases 84
Real gas
  fugacity of a — in a mixture 83
  partial molar volume of a — 84
  pure — 75
Reduced variables 79
Reduction potential 175
Reference state for entropy 101
Refrigerator 31, 33
Relation
  between activity coefficients in binary systems 160
  between C_p and C_V 51-52
  Maxwell's —s 46-48
  —s between partial molar quantities 64
Reverse process and the sign of work 12
Reversible
  adiabatic process of an ideal gas 29
  isothermal expansion or compression of an ideal gas 26
  process 7, 19

Rule
  lever — 143, 144, 159
  Lewis-Randall — 87
  phase — 3, 91, 133
  phase — and electrolytes 170
Schwarz theorem 7, 46, 52
Second law of thermodynamics
  Spontaneous processes and the — 57
  — 17-37
  Application to chemical equilibrium 119
  Kelvin formulation of the — 17
Selection of the standard state pressure 69
Sign convention for energy exchanges 9
Simultaneous
  independent — reactions 128
  reactions 131
Solute
  — 147, 162
  effect of a — on the boiling temperature 150
  freezing point lowering by a — 147
Solution
  ideal — 140
  isothermal diagram 141
  non ideal — 153
  perfect and ideal — 139-152
  perfect — 141
  real — 153
Solvent
  — 147, 162
  elevation of the boiling temperature of a — by a solute 150
  lowering of the freezing point of a — by a solute 147
Spontaneous
  evolution of a system 57
  process 20, 75
  transfer of species between phases 90
Standard
  chemical potential 69
  chemical potential in liquids 157
  chemical potential using convention I 157

chemical potential using convention II 163
entropy 107
equilibrium constant 123
Gibbs energy of formation 107
heat capacity at constant pressure 107
molar heat capacity at constant pressure 107
variables of reaction 105
Standard state
— 106
of a real gas 76
on the molality scale 166
pressure 69, 105
temperature 105
State
corresponding —s of real gases 80
equation of — 8
function 2, 3, 15, 89
standard — 106
standard — and convention I for the activity coefficient 157
standard — and convention II for the activity coefficient 162
variable of — 3
Stirling cycle 36
Stoichiometric coefficients 103
Sublimation 98
System
at equilibrium 1
binary — 158
closed adiabatic — 20
closed — 1, 40-41
heterogeneous — 1, 126
heterogeneous — and the law of mass action 126
homogeneous — 1
in contact with one thermal reservoir 17, 21, 53-56
in contact with several thermal reservoirs 22
in contact with two heat sources 17
irreversible process in a — 20

isolated — 1
open — 1, 42-46, 90
without chemical reaction 89
Temperature
critical — 78
effect of — on chemical equilibria 136
effect of — on the activity coefficient 155
effect of — on the equilibrium constant 134
inversion — 82
standard state — 105
thermodynamic — 18
Thermal
coefficients 4
equilibrium 1, 2
equilibrium and entropy 24
expansion coefficient 4
machines 31
Thermal reservoir
definition of a — 1
entropy change of a — 21, 26, 54, 75
system in contact with one — 17, 21, 53
system in contact with several —s 22
Thermodynamic equilibrium constant 123, 127
Thermodynamic temperature 18
Third Law of thermodynamics 101
Transfer
heat — 24
spontaneous — of species 90
Triple point 19, 99
Van der Waals
coefficients 79
equation of state 78
Vapor pressure
— 95, 140
effect of an inert gas on — 95
effect of temperature on — 97
of an ideal solution 141
Vaporization curve 98, 141, 144, 161

Index

Variable
    auxiliary state — 3, 39
    auxiliary — 39
    excess — of mixing 154
    extensive — 2, 39
    intensive — 2, 39
    natural —s 42
    of mixing 154
    of mixing for real gases 84
    of reaction 104, 105
    of state 3
    reduced —s 79
    standard —s of formation 107
    standard —s of reaction 105
Variance of a system 91, 133
Variation of heat capacities with volume and pressure 52
Virial coefficients and equation 77
Volume
    effect of — on chemical equilibria 137
    of mixing 145
    of reaction 105
    partial molar — of a real gas 84
    partial molar — of an ideal gas 72
    work due to — change 10
Work
    any form of — 113
    due to volume change 10
    electrical — 12
    mechanical — 9
    other than volume — 114
    sign convention for — 9
    various expressions for — 13
Zeroth law of thermodynamics 2

www.ingramcontent.com/pod-product-compliance
Lightning Source LLC
Chambersburg PA
CBHW030937180526
45163CB00002B/597